Lactate

Physiologic, Methodologic and Pathologic Approach

Edited by
P.R. Moret, J. Weber
J.-Cl. Haissly, H. Denolin

With 73 Figures and 12 Tables

Springer-Verlag
Berlin Heidelberg New York 1980

Professor P.R. Moret, Division de Cardiologie, Département de Médecine, Hôpital Cantonal, CH–1211 Genève 4

Dr. J. Weber, P.D., Médecin-chef du Service de Médecine, Hôpital de Montreux, CH-1820 Montreux

Dr. J.-Cl. Haissly, La Levratte 26, CH–1260 Nyon

Professor H. Denolin, Président de la Société Européenne de Cardiologie, 178 Avenue Winston Churchill, B–1180 Bruxelles

ISBN-13:978-3-540-09829-4 e-ISBN-13:978-3-642-67525-6
DOI: 10.1007/978-3-642-67525-6

Library of Congress Cataloging in Publication Data
Main entry under title:
Lactate physiologic, methodologic, and
pathologic approach.
Bibliography: p.
Includes index.
1. Lactic acid–Metabolism–Congresses.
2. Lactic acid–Analysis–Congresses. 3. Heart–
Diseases -Congresses. 4. Exercise–Physiological
aspects–Congresses. I. Moret, Pierre R.
[DNLM: 1. Lactates. QU98 L151]
QP801.L3L32 599.01'2 79-28452
ISBN-13:978-3-540-09829-4

2127/3321-543210

Contents

List of Contributors

BAUDRAS, A., Centre de Recherche de Biochemie et de Génétique Cellulaires du CNRS 118, Route de Narbonne, 31077-Toulouse Cedex

BERGER, W., Department of Medicine, University Hospital, Spitalstr. 21, CH-4031 Basel

CERRETELLI, P., Centro Studi di Fisiologia del Lavoro Muscolare del CNR, Instituto di Fisiologia Umana III, Università di Milano, Via Mangiagalli 32, 20133-Milano

de COSTER, A., Département des voies respiratoires, Service de Médecine Interne, Hôpital Universitaire Saint-Pierre, 322, rue Haute, 1000 Bruxelles

DAWSON, M.J., Department of Physiology, University College London, Gower Street, London WC IE 6BT

DEGRE, S., Département de Cardiologie, Hôpital Universitaire St. Pierre, 322 rue Haute, 1000 Bruxelles

DENOLIN, H., Département de Cardiologie, Hôpital Universitaire St. Pierre, 322, Rue Haute, 1000 Bruxelles

DURAND, J., Centre Marie-Lannelongue, 133, Avenue de la Résistance, 92350 Le Plessis Robinson

ENRICO, J.F., Intensive Care Unit, Medical Department, and Division of Pathophysiology, University Hospital, 1011 Lausanne

FREMINET, A., Institut National de la Santé et de la Recherche Médicale, Unité 27, and Institut d'Anesthésiologie, 42, rue Desbassayns de Richemont, 92150 Suresnes

GADIAN, D., Department of Biochemistry, University of Oxford, South Parks Road, Oxford OX1 3 Qr

HARRIS, P., Cardiothoracic Institute, University of London, 2, Beaumont Street, London, WIN 2DX

HEARSE, D.J., Myocardial Metabolism Research Laboratories, The Rayne Institute, St. Thomas' Hospital, London S.E.1

HERMANSEN, Lars, Institute of Work Physiology, University of Oslo, Oslo

HOWALD, H., Research Institute of the Swiss School for Physical Education and Sports, 2532 Magglingen

KELLER, U., Department of Medicine, University Hospital, Spitalstr. 21, CH-4031 Basel

KRÄHENBÜHL, B., Unité d'Angiologie, Policlinique Universitaire de Médecine, Hôpital Cantonal, CH-1211 Genève 4

KUGLER, G., Division of Cardioangiology, Department of Medicine, University of Hamburg, Martinstraße 52, D-2000 Hamburg 20

LECLERG, L., Institut National de la Santé et de la Recherche Médicale, Unité 27, and Institut d'Anesthésiologie, 42, rue Desbassayns de Richemont, 92150 Suresnes

MESSIN, R., Département de Cardiologie, Hôpital Universitaire Saint-Pierre, 322 rue Haute, Bruxelles

MORET, P.R., Division of Cardiology, University Hospital, CH-1211 Geneva

MUNZ, E., Biochemica Werk Tutzing, Bahnhofstr. 5, D-8132 Tutzing

OPIE, L.H., Fisiologica Clinica CNR, l'Universita degli Studi di Pisa, Via Savi 8, 56.100 Pisa

PERRET, C., Intensive Care Unit, Medical Department, and Division of Pathophysiology, University Hospital, 1011 Lausanne

PIIPER, J., Abteilung Physiologie, Max-Planck-Institut für experimentelle Medizin, D-3400 Göttingen

POOLE-WILSON, P.A., Cardiothoracic Institute University of London, 2, Beaumont Street, London, WIN 2 DX

POYART, C., Institut National de la Santé et de la Recherche Médicale, Unité 27, and Institut d'Anesthésiologie, 42, rue Desbassayns de Richemont, 92150 Suresnes

REMME, W.J., Department of Cardiovascular Research, Thoraxcentrum, Erasmus University Rotterdam, Postbus 1738, Rotterdam

ROSSEL, C.F., University Hospital, Geneva

STAM, H., Department of Cardiovascular Research, Thoraxcentrum, Erasmus University Rotterdam, Postbus 1738, Rotterdam

VAAGE, O., Institute of Work Physiology, University of Oslo, Oslo

VERDOUW, P.D., Department of Cardiovascular Research, Thoraxcentrum, Erasmus University Rotterdam, Postbus 1738, Rotterdam

WEBER, J., Montreux Hospital, CH-1820 Montreux

WILKIE, D.R., Department of Physiology, University College London, Gower Street, London WC IE 6 BT, England

WUHRMANN, H.R., F. Hoffmann-La-Roche & Co. A.G., Department of Bio-Electronics, CH-4002 Basel

Preface

This book covers the principal reports and discussions which have been presented during the "Symposium on Anaerobic Metabolism and Lactic Acid" which was held in Anzère (Wallis, Switzerland) from 19th till 20th January 1978 under the patronage of the European Society of Cardiology. As Dr. Wilkie and Hermansen mentioned in their paper, Berzelius was the first one who recognized in 1807 "the existence of free lactic acid in the muscular fluid and thought that the amount of free lactic acid in a muscle is proportional to the extent to which it has been previously exercised". Since, many controversies have appeared, not as far as for its presence in the muscle, but as for its relations to oxygen debt or fatigue, to its physiological part, to its metabolism (oxydation or conversion to glycogen in situ), or to its pathological significance in clinical conditions.

In this symposium, physiologists and biochemists tried to resolve some of the basic problems, they defined terminology and introduced new facts. Methods to determine the lactic acid or lactate have been reviewed. Lactic acid or lactate in various clinical conditions were the objectives of many collected reports in the last part of the symposium. Clinicians, who are interested in Pneumology, Diabetology, Hepatology, Angiology, Intensive Medicine or Sport Medicine, and particularly cardiologists will find precious information.

The organizing committee thanks every reporter, specially the Chairmen, and everyone who took an active part in the preparation or realization of this symposium. We hope that the present book will bring everyone useful elements of information.

P.R. Moret
J. Weber

Section I
Physiology

Opening Remarks

D.R. Wilkie

Firstly, I wish to express my thanks to the Société Européenne de Cardiologie and to Dr. Moret personally for organising this Symposium on "Anaerobic Metabolism and Lactic Acid", and for inviting me to participate in it.

It was in 1807 that the great Berzelius "recognised the existence of free lactic acid in the muscular fluid" and "thought that he had convinced himself that the amount of free lactic acid in a muscle is proportional to the extent to which it has been previously exercised" (1). It might provide a challenging theme for this Meeting to take a hard look at our progress since 1807, with the aim of sorting out the facts that can be depended on as established from the others about which there is still disagreement. In these controversial areas we must try to discover whether the controversy is about facts, about concepts or merely about the use of words. In the last two cases, we may be able to resolve matters here and now at this Symposium. If experimental facts are in question we should decide what key experiments should be done to resolve the matters. This amounts to an ambitious programme for two-and-a-half days, so without further ado, we had better begin.

References

1. BERZELIUS, J. cited by LEHMANN, C.F.: Physiological chemistry, 2nd ed., vol. I, p. 98. London: Harrison

See also the article by Hermansen in this volume.

Chapter 1

Lactate Metabolism and Cardiac Muscle

L.H. Opie

Introduction

Lactate has two established functions in the metabolism of the heart. First, during aerobic metabolism lactate is a normal myocardial fuel, which during certain conditions (e.g., exercise) can account for the major part of the myocardial oxygen consumption. A supply of lactate appears to depend on competition with other myocardial fuels, especially free fatty acids. Secondly, during anaerobic metabolism, lactate production is viewed as part of the Pasteur effect, whereby anaerobiosis accelerates glucose uptake and glycolysis, and pyruvate, instead of entering the citrate cycle, forms lactate.

Lactate Versus Lactic Acid

Although these two terms are frequently used interchangeably, we should bear in mind that: (1) enzymatic methods measure lactate and not lactic acid; (2) there may be different rates of production of lactate and protons, with the latter actually derived from ATP breakdown; (3) there may be different rates of transport of lactate and lactic acid across the cell membrane (1); and finally (4) there may be different rates of disposal of lactate and protons (see communication by Piper, this volume). For the above reasons, it seems that most communications at this meeting should be referring to lactate and not to lactic acid. An exception is the situation in which lactate and protons are measured and reported separately, e.g., in the clinical condition of lactic acidosis, when the lactate concentration rises and the blood pH falls. Another exception is when lactic acid is allowed to ionise freely and such ionisation takes place in conditions completely buffering the protons formed; this can occur in normal muscle if small amounts of lactic acid have been added (e.g., to homogenised muscle in vitro). But if lactic acid is added in large amounts the tissue pH will fall. In the ischaemic myocardium, lactate and protons both increase but they derive from different immediate sources (2).

Lactate and Ischaemia

During ischaemia, lactate plays a new role; that of metabolic regulator. Thus lactate can cause (1) feed-back inhibition of glycolysis at the level of glyceraldehyde 3-phosphate dehydrogenase, which should be distinguished from the inhibition of glycolysis caused at the phosphofructokinase level by protons (3); (2) shortening of the action potential duration (4); and (3) increased enzyme release and inefficiency of coronary-ligated isolated perfused rat hearts (Table 1). Effects (2) and (3) are measured against reference values with glucose. Thus accumulation of lactate may be a mechanism promoting ischaemic damage and possibly provoking arrhythmias (by shortening of the action potential duration).

Table 1. Effects of lactate compared with glucose and free fatty acid on release of lactate dehydrogenase, mechanical performance, and efficiency of isolated rat heart with coronary artery ligation. (7)

Group	Substrate (mmol/litre)	Number of hearts	LDH (U/g/h)	Efficiency Joules/ml O_2
A	Palmitate 1.5 Albumin 0.25	22	27.6 ± 2.3 (22)	3.91 ± 0.08 (12)
B	Glucose 11 Albumin 0.25	7	6.7 ± 0.3 (7)	4.93 ± 0.14 (7)
	P vs A		$\ll$ 0.001	$\ll$ 0.001
C	Lactate 5 Albumin 0.25	8	12.6 ± 2.3 (8)	4.63 ± 0.16 (8)
	P vs A		< 0.005	< 0.001
	P vs B		< 0.05	NS

Effect of lactate was shown in increasing enzyme release when compared with glucose, and in increasing efficiency when compared with palmitate. $P < 0.025$ for comparisons of means ± SEM for all readings for each heart (4.93 ± 0.09, n = 35 vs 4.46 ± 0.09, n = 40) when efficiency of lactate- and glucose-treated hearts are compared. This degree of change in efficiency could be expected from the difference in respiratory quotient: glucose 3.15 and lactate 3.00

Clinical Application

These undesirable effects of lactate may explain the association between high blood lactate values and increased mortality in patients with acute myocardial infarction (5). Thus a vicious circle could be proposed, as follows:

large infarcts → increased catecholamine secretion → increased blood lactate → increased ischaemic damage and arrhythmias → large infarcts.

Reservations

Reservations to the above conclusions are as follows: First, shortening of the action potential duration does not necessarily prove that lactate is arrhythmogenic, and further experiments are required. Secondly, the early release of enzyme within 60 min from an isolated heart preparation does not prove that lactate would worsen the infarction process, which extends over many hours in experimental animals such as dogs and in man; free fatty acids are thought to participate in a metabolic vicious-circle mechanism (6) but at the same time it is known that free fatty acids and lactate compete for the myocardial oxygen supply. No details about this competition are known in conditions of ischaemia, but it is possible to visualise that lactate might be less harmful to the ischaemic myocardium than free fatty acids, as is indeed supported by experimental evidence (Table 1). It may be that promotion of lactate metabolism by reducing free fatty acid metabolism might actually have some beneficial effect. Finally, it should be noted that at high blood lactate concentrations there is a very low concentration of lactic acid, which is in the micromolar range when lactate is in the millimolar range, and such lactic acid may cross the cell membrane more freely than does lactate (I); once inside the cell, the lactic acid could dissociate freely to produce protons. Thus some of the external lactate effects could be due to an internal acidosis.

Summary

The effects of lactate are not necessarily the same as those of lactic acid. Lactate by itself, in the absence of added protons, has important effects on the heart, which may lead to aggravation of myocardial ischaemia. Reservations to these conclusions are outlined.

References

1. COHEN, R.D., ILES, R.A.: Lactic acidosis: some physiological and clinical considerations. Clin. Sci. Mol. Med. 53, 405-410 (1977)
2. GEVERS, W.: Generation of protons by metabolic processes in heart cells. J. Mol. Cell. Cardiol. 9, 867-874 (1977)
3. ROVETTO, M.J., LAMBERTON, W.F., NEELY, J.R.: Mechanisms of glycolytic inhibition in ischemic rat hearts. Circ. Res. 37, 742-751 (1975)
4. OPIE, L.H., NATHAN, D., LUBBE, W.F.: Biochemical aspects of arrhythmogenesis and ventricular fibrillation. Am. J. Cardiol. to be published
5. JAHRMÄRKER, H., RACKWITZ, R., HAIDER, M.: Blutlaktat als Parameter zur Überwachung bei akutem Myokardinfarkt. In: The first 24 hours in myocardial infarction. Kaindl, F., Pachinger, O., Probst, P. (eds.), pp. 51-53. Baden-Baden New York_ Witzstrock 1977
6. OPIE, L.H., Tansey, M.J., KENNELLY, B.M.: Proposed metabolic vicious circle in patients with large myocardial infarcts and high plasma free fatty acid concentrations. Lancet 2, 890-892 (1977)
7. OPIE, L.H., BRICKNELL, O.L.: Substrate-induced enzyme release and inefficiency of work in isolated rat hearts with coronary artery ligation. Effects of fatty acid, lactate, glucose and glucose-insulin-potassium. To be published

Discussion

Dr. Opie mentioned in his talk that we should be careful to distinguish between the production by muscle of lactate and of lactic acid. This view appears to be mistaken, since in the presence of the buffers found in muscle, which hold the pH at around 7, 99.92% of the lactic acid formed is very rapidly converted to lactate.

$$HL \rightleftharpoons H^+ + L^- \qquad pK = 3.9,\ K = 1.26 \times 10^{-4}$$

$$[HL] / [L^-] = 10^{-7}/1.26 \times 10^4 = 7.9 \times 10^{-4} = 0.08\%$$

The same criticism was made independently by Professor Piiper and Dr. Poole-Wilson.

A second criticism concerned several of the chemical equations shown by Dr. Opie to illustrate his arguments about the production and absorption of protons in various biochemical reactions. I think that by showing both the stoichiometric coefficients and the changes as whole numbers, as in

$$ATP^{4-} \longrightarrow ADP^{3-} + P_i^{\,2-} + H^{1+} \qquad (1)$$

at pH 7.2, Dr. Opie has been led into some wrong conclusions. Although Equation 1 is typical of the way in which reactions are portrayed in many textbooks, it is in fact incorrect. Reactions in which one or more of the reactants has a pK reasonably close to the pH at which the reaction is occurring are more complex, and should be written with fractional charge numbers or, preferably, fractional stoichiometric coefficients.

Let us start with a simple example, the hydrolysis of phosphocreatine (PCr) to creatine (Cr) and inorganic orthophosphate (P_i). This is directly relevant to the articles by Dawson et al. and by Piiper in this volume. The reagents with which we are concerned have the following ionisation characteristics:

$$H_3PO_4^{0} \rightleftharpoons H^+ + H_2PO_4^{1-} \rightleftharpoons H^+ + HPO_4^{2-} \rightleftharpoons H^+ + PO_4^{3-} \qquad (2)$$

pK = 2   6.78   11.7

$$PCr^{1+} \rightleftharpoons H^+ + PCr^{0} \rightleftharpoons H^+ + PCr^{1-} \rightleftharpoons H^+ + PCr^{2-} \rightleftharpoons H^+ + PCr^{3-} \qquad (3)$$

pK = 2.0   2.6   4.5   12-13

and

$$Cr^{1+} \rightleftharpoons H^+ + Cr^{0} \rightleftharpoons H^+ + Cr^{1-} \qquad (4)$$

pK = 2.6   12-13

The t+ forms of PCr and Cr result from strong (pK = 12-13) protonation of one of the N atoms. Only P_i has a pK close to the reaction pH, 7.2. At this pH

$$\frac{HPO_4^{2-}}{H_2PO_4^{1-}} = \frac{10^{-6.78}}{10^{-7.2}} = 2{,}6302 \qquad (5)$$

That is, if we have one mole of P_i in all forms, 0.7245 mole of it is present as HPO_4^{2-} and 0.2755 mole as $H_2PO_4^{1-}$. (I have chosen in this section to give a numerical treatment of one example. A few minutes' work with pocket calculator can often give much more insight than does the original algebra.) Thus the whole equation can be written:-

$$PCr^{2-} + H_2O \rightleftharpoons Cr^{0} + 0.7245\ HPO_4^{2-} + 0.2755\ H_2PO_4^{1-} - 0.2755\ H^+ \qquad (6)$$

or alternatively

$$PCr^{2-} + H_2O \rightleftharpoons Cr^{0} + P_i^{1.7245-} - 0.2755\ H^+. \qquad (7)$$

One has a choice between the use of fractional stoichiometric coefficients and that of fractional charges. The former, though longer, is probably clearer. In either case the source and amount of the absorption of protons from the buffer system (alkalinisation) is made very obvious.

The reaction shown in Equation 1 is more complex, because both ATP and ADP have pK values in the region of 7.2.

$$ATP^{1+} \rightleftharpoons 3H^+ + ATP^{2-} \rightleftharpoons H^+ + ATP^{3-} \rightleftharpoons H^+ + ATP^{4-} \qquad (8)$$

pK = 2   4.06   6.95

$$ADP^{1+} \rightleftharpoons 2H^+ + ADP^{1-} \rightleftharpoons H^+ + ADP^{2-} \rightleftharpoons H^+ + ADP^{3-} \qquad (9)$$

pK = 2   3.93   6.88

The positively charged forms arise from weak (pK ≅ 4) protonation of the adenine ring. By exactly similar reasoning to that employed above it is then easy to show that at pH 7.2 reaction (1) should be written:

$$0.36\ ATP^{3-} + 0.64\ ATP^{4-} + H_2O$$

$$= 0.3237\ ADP^{2-} + 0.6763\ ADP^{3-} + 0.7245\ HPO_4^{2-} + 0.2755\ H_2PO_4^{1-} + 0.7608\ H^+,$$

showing that in this reaction 0.7608 mole of protons is produced per mole of ATP hydrolysed. The fact that all these stoichiometric coefficients are fractions that vary with pH should clearly be taken into account in discussing problems such as those posed by Dr. Opie.

D.L. Wilkie

Discussion

1. Lactate versus lactic acid.
   The data given by Professor Wilkie for the dissociation of lactic acid are generally accepted. Such data do not in any way negate the thesis that neutral lactate, added in a variety of systems, has effects on cardiac function. If lactic acid were added, as Professor Wilkie rightly points out, there would be the combined effects of the lactate plus those of protons. In the setting of myocardial ischaemia, where the intracellular pH falls, the effects of lactate plus protons may well be different from those of lactate.

2. The chemical equations shown by me in my talk were based on those given by Gevers. I am no expert in the matter but as far as I can see, the principles propounded by Gevers are in no way altered by the analysis that Professor Wilkie makes. As Gevers states, the exact number of protons produced per mole of ATP hydrolysed would vary according to the pH and to other factors, including magnesium binding. The different effects of various pH values were stressed by Gevers. The important point that he makes is that the immediate source of protons during anaerobic glycolysis is the breakdown of ATP and that continued production of protons depends on continued turnover of ATP. Thus the protons originally derived from glucose are incorporated into ATP and then regenerated. Lactic acid is not made directly. I believe that these analyses by Gevers have theoretical validity, although it remains to be shown that they have practical importance.

References

GEVERS, W.: Generation of protons by metabolic processes in heart cells. J. Mol. Cell. Cardiol. 9, 867-874 (1977)

Chapter 2

Studies of Fatigue in Intact Anaerobic Living Muscle with Nuclear Magnetic Resonance (^{31}P NMR): Present Results and Future Possibilities

M.J. Dawson, D.R. Wilkie and D. Gadian

We have already described (1) how the noninjurious radiofrequency technique of ^{31}P NMR can be applied to isolated muscles during rest, contraction, and recovery. However, it seems that the physical principles underlying the method are not universally understood and that it would be helpful to explain this aspect briefly, so that the potentialities and the limitations of NMR may be more readily appreciated.

Elementary NMR

The use of NMR to measure tissue metabolites imposes constraints quite different from those encountered in conventional physiological or NMR experiments. To appreciate these constraints it is necessary to know something about the physical principles of NMR. All atomic nuclei have positive charge and mass but only some of them, including ^{1}H, ^{2}H, and ^{31}P, possess spin. The rotating charge causes the nucleus to behave like a tiny magnet and the spinning mass endows it with the properties of a gyroscope. Thus when an external magnetic field is applied, the nuclear magnets do not line up with it in the same way as nonspinning magnets would. Instead, they precess like a child's top around the direction of the field at a characteristic frequency, the Larmor frequency ν_L, which is directly proportional to the magnetic flux density B_o. For ^{31}P nuclei, $\nu_L = 1.723 \times 10^7 B_o$.

Thus in the (vertical) 7.5 Tesla field of the superconducting Oxford instrument, designed by Hoult and Richards (2) the nuclei precess at 129.2 MHz. If a horizontal magnetic field oscillating at this frequency is applied, the nuclei absorb energy and will continue to radiate after the exciting field is switched off at a frequency proportional to the local magnetic field experienced by the mucleus. This differs very slightly from the applied magnetic field B_o because of the effects of local circulating electrons, neighbouring magnetic nuclei, etc., which are characteristic of different types of compound. The resulting frequency shift, the chemical shift, can be detected and used to identify the different phosphorus compounds present. The chemical shift is usually expressed in parts per million (parts/10 6).

Unfortunately NMR is an insensitive technique: in our experiments only a weak signal, hardly visible in the noise, is emitted following the 40-μs pulse at 129.2 MHz used for excitation. This makes it necessary to average many signals or "Scans" to obtain useful spectra. Although the signal decays within about 20 ms ($T_2 \cong 10$ ms) the pulse can be repeated only every 2 s because it takes a relatively long time (determined by T_2, the spin-lattice relaxation time) for the ^{31}P nuclei to revert to their unexcited state. As a result, the accumulation of a spectrum can take several hours.

In studying muscular contraction and recovery the inherent slowness of the method is overcome by stimulating repeatedly and averaging the signals during present time intervals phased with the stimulus (see Fig. 2).

Physiological Techniques

The physiological problem is the counterpart of the physical situation. The choice of experimental preparation thus involves a compromise; thick muscles would give large signals but only thin ones can be adequately oxygenated by diffusion. They should, if possible, function at low temperature. Accordingly, most of our experiments were performed on frog muscles from *Rana temporaria* at $4^{o}C$. In our present experiments on fatigue we have deliberately worked under anaerobic conditions produced by N_2 and NaCN, so as to be able to work with large frog gastrocnemii and thus obtain adequate spectra (see Fig. 2) in a reasonably short time.

The spectra from resting, oxygenated frog and toad muscles are shown in Fig. 1. The positions of the peaks identify each compound, and

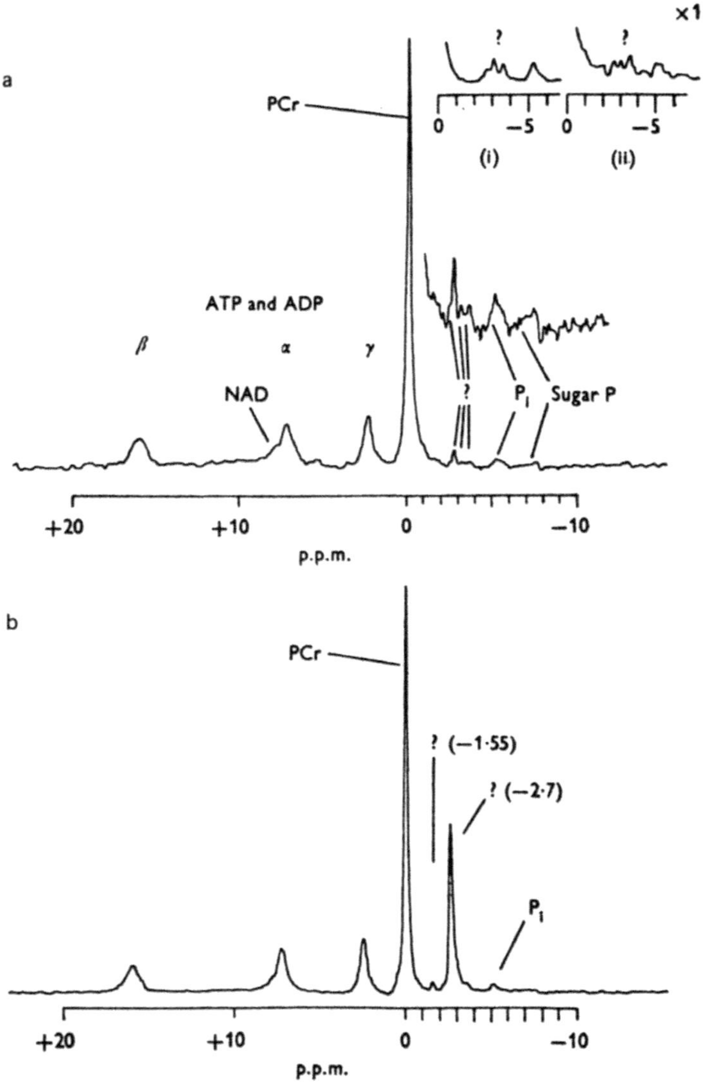

Fig. 1 a and b. Spectra from resting muscles at 4°C. Ordinates, signal, abscissae, chemical shift in p/10^6; the frequency increases from left to right, the total span representing 5 kHz. a Four frog sartorri, May 19, 1976. Lowest line: Average of 10 000 scans at 2-s intervals, with peaks assigned as described in text. The short length of the record on the right-hand side of the PCr peak has been enlarged 4 x vertically to show fine detail. Note that of the three unidentified peaks at -2·7, -3·05 and -3·6 p/10^6, only the left-hand one is conspicuous. The inset figures show how these peaks vary in size, though not in position. (i) 16 November 1975; (ii) 29 April 1976. Two toad gastrocnemii, 20 May, 1976. Average of 6 000 scans at 4-s intervals. M.J. Dawson, et al.: J. Physiol. 267: 703-735, 1977 the area under the peak is proportional to the amount of compound present. Note the small amount of inorganic phosphate (Pi) present in these intact living muscles. The position of this Pi peak shifts according to the pH: the observed position thus provides an excellent measure of the intracellular pH. The positions of the ATP peaks, especially those arising from the terminal β and γ phosphorus atoms, are changed when Mg^{2+} is bound: in living muscle almost all the ATP is indeed present as MgATP, which is important since only MgATP is a useful substrate for actomyosin ATPase and for other enzyme reactions. The region marked "Sugar P" indicates the zone occupied by the hexose and triose phosphates of the glycolytic pathway. Finally, the peaks marked "?" represent compounds whose functional significance is at present unknown, but which may be present in quite large concentrations (1, 3, 4).

In our present experiments we have used ^{31}P NMR to monitor the metabolic changes in anaerobic frog gastrocnemius muscles undergoing repeated fatiguing isometric contractions at $4^{\circ}C$. A typical experiment is illustrated in Fig. 2. The relationships between decline in force development and changes in all of the following were determined: (a) concentrations of phosphorus compounds, creatine, and H^+; (b) free-energy change for ATP hydrolysis; (c) rate of lactic acid (LA) production; and (d) total ATP hydrolysis. Of these, the concentrations of H^+ and of the phosphorus compounds can be determined directly by ^{31}P NMR, as explained above. The other quantities were estimated on the basis of current knowledge of muscle biochemistry together with the ^{31}P NMR results. After 2 h poisoning with 2 m$\underline{M}$ NaCN, muscles were tetanised for 1 s/20 s, 1 s/60 s, or 5 s/300 s. We performed two experiments of each type, lasting from 18 to 92 min. The different patterns of stimulation used in these experiments ensured that there were differences in all of the following: (1) the total duration of the experiment; (2) the total number of contractions; and (3) the total period for which the muscles were stimulated (i.e., the total number of contractions multiplied by duration of each contraction). When force development in individual experiments was plotted against these three parameters, the results fell into groups according to the different patterns of stimulation. In contrast, the relationship between force development and concentration of each of the measured chemical substances is completely independent of the pattern of stimulation. We conclude that the decline in isometric force development is more likely to

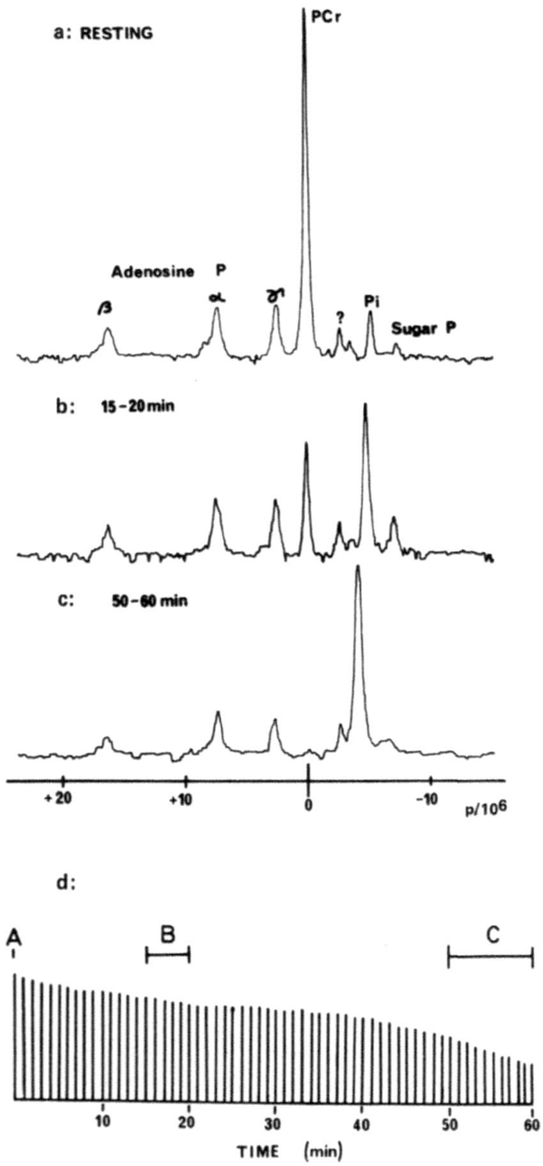

Fig. 2 a-d. Results from a single experiment in which a pair of anaerobic frog gastrocnemii (total wt = 773 mg) were stimulated 1 s/60 s. Panels a, b, and c are examples of the ^{31}P-NMR spectra obtained, with peaks identified in spectrum a. The multiple peak centred at -3.0 p/10^6 and marked? represents three compounds of unknown function. Two of these have been identified as glycero-phosphoryl choline[4] and serine-ethanolamine phosphodiester[22]. The peak labelled Sugar P at -7.5 p/10^6 represents combined resonances from several hexose and triose phosphates; AMP and IMP appear in the same general region. We found no consistent or substantial changes in the? or Sugar P peaks, so they are neglected in the body of the text. Spectrum a was accumulated for 10 min before the first stimulation and c is the sum of two spectra accumulated in separate 5-min bins. Panel d is a drawing, indicating the average value of each force response; the initial force was 238 mN · mm^{-2}

be linked to the levels of the measured biochemical substances than to any changes in excitatory conduction. Conduction changes could be responsible for the decline in isometric force only if they themselves are somehow linked to biochemical changes; such a mechanism for fatigue has previously been suggested. (5)

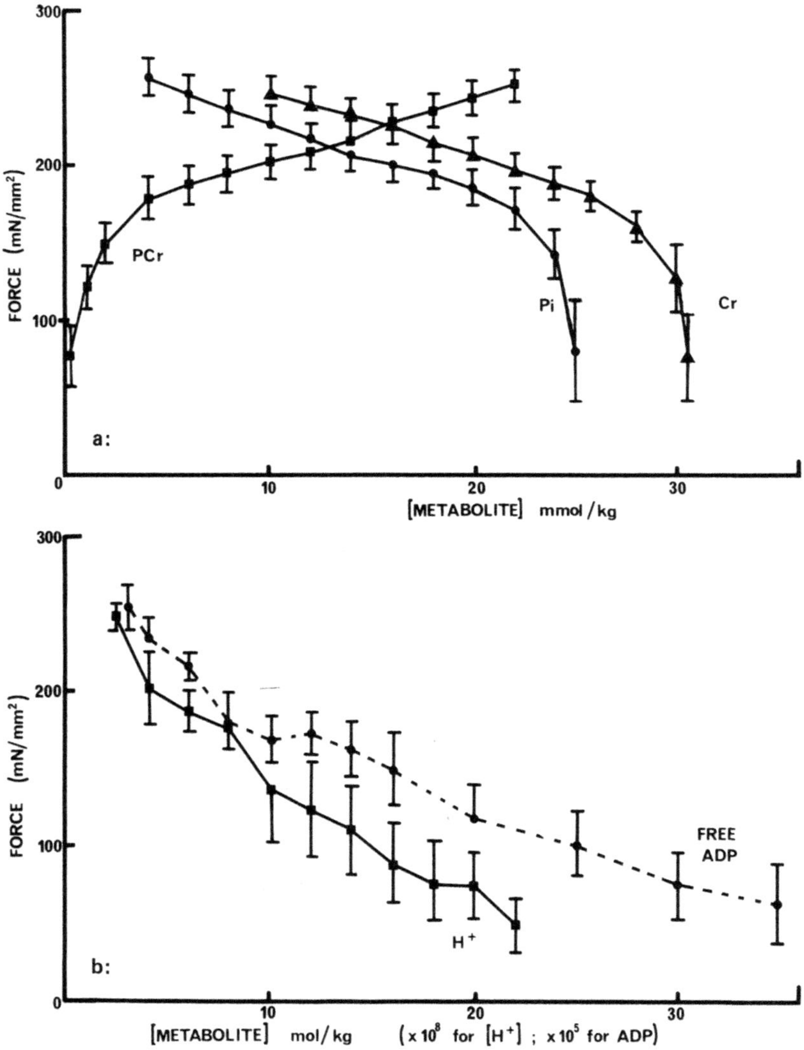

Fig. 3. Force developed as a function of metabolite levels; averaged results of six experiments. Spectra were accumulated for 5-min periods if the muscles were stimulated for 1 s/60 s or 5 s/5 min, and for 2-min periods if the stimulation was for 1 s/20 s. In the 5 s/5 min experiments the average force developed by each contraction was related to the metabolite levels indicated by the spectrum preceding that contraction. For the 1-s stimulation experiments the force developed by the contraction nearest the middle of the spectral-averaging period was related to the metabolite levels. The force developed at particular metabolite levels was obtained by linear interpolation between the data points, so as to obtain the standard errors (shown by vertical bars)

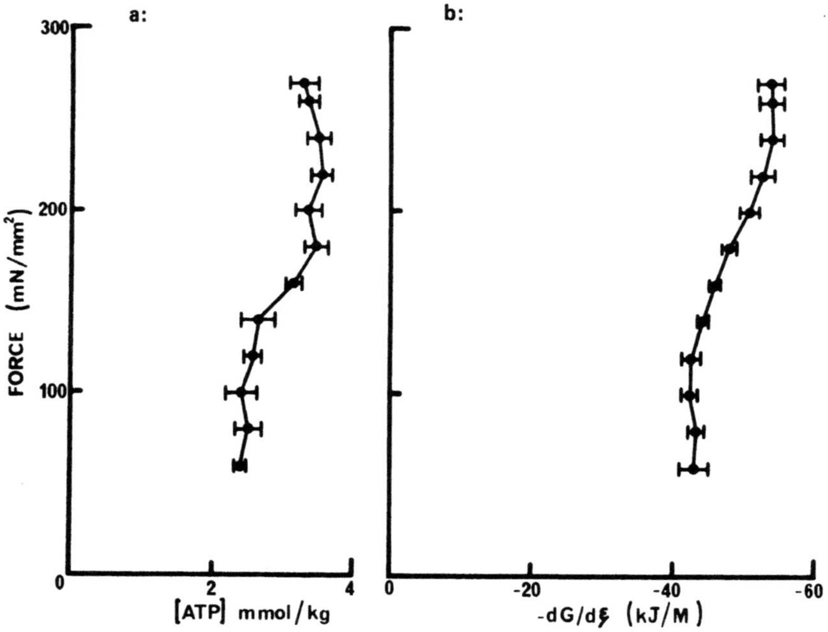

Fig. 4 a and b. Force as a function of [ATP] (a) and affinity (b). The data were obtained as explained in the legend to Fig. 2. The horizontal bars represent standard errors

Since there was no discernible effect of the pattern of stimulation on the biochemical results we have averaged all six experiments after grouping them for statistical purposes by linear interpolation.

Results

1) Metabolite and H^+ Levels.

In addition to measuring [ATP], [PCr], [Pi], and [H^+] directly, we calculated [creatine] from the observed changes in [PCr], and [free ADP] from the creatine phosphotransferase equilibrium. A close correlation was found between force development and the levels of each of these metabolites; these relationships were completely independent of the pattern of stimulation (see Figs. 3 and 4). The [ATP] dropped by about 1 mM and [PCr] dropped nearly to zero, while force declined to 20% of its initial value. The relations between force and [PCr], [Pi], or [Cr] were clearly nonlinear, thus running counter to the results reported by Spande and Schottelius (6), but the relations between force and [free ADP] or [H^+] may be linear.

2) Free-Energy Change.

Alberty's method (7) was used to calculate the free-energy change ($A = -dG/d\xi$), often miscalled $-\Delta G$ (8), for ATP hydrolysis. The decline in $-dG/d\xi$ was similar in all the experiments and was small (see Fig. 4B). The significance of $-dG/d\xi$ is that it measures the maximum work theoretically available per mole of ATP hydrolysed.

Glycolysis and ATP Turnover

We have so far been concerned with relating force development to levels of affinity or of metabolites. However, the situation in muscle is dynamic, the metabolite levels reflecting the balance between the rate at which ATP is being hydrolysed and the rate at which it is being resynthesised. Comparison of the changes in rates of these reactions with the decline in force development yields additional information concerning the mechanism of fatigue.

Under anaerobic conditions resynthesis of ATP occurs mainly from glycolysis and from the breakdown of PCr. The quantity of lactic acid (LA) produced can be estimated from the pH changes and the buffering capacity of frog muscle. Other reactions known to be occurring in contracting muscle have much smaller effects on $[H^+]$ and have therefore been neglected. It is known that LA diffuses only very slowly out of frog muscle; the results of Mainwood and co-workers (9) on fatigued frog muscle indicate maximum rates of LA diffusion into the medium of only a few micromoles per kilogram per hour. We have accordingly neglected LA diffusion from the muscle in the following calculations.

Over the range of pH that concerns us (6.5-7.5), the concentrations of undissociated LA (pK = 3.9) and of H^+ are both negligible compared with the concentrations (> 1 m*M*) that concern us here. LA^- production can thus be estimated from the increase in the BH^+ forms of the buffers present. For one buffer (B, dissociation constant K) whose total concentration $([BH^+] + [B]) = C$

$$[BH^+] = C\,[H^+]/(K + [H^+]) \qquad (5)$$

In muscle the production of LA^- can therefore be calculated from:

$$\Delta[LA^-] = \Delta\left\{\frac{[\text{protein Hist}]\,[H^+]}{K_{hist} + [H^+]} + \frac{[\text{carnosine}]\,[H^+]}{K_{carn} + [H^+]} + \frac{[Pi]\,[H^+]}{K_{Pi} + [H^+]}\right\} \quad (6)$$

$$= \Delta\left\{\frac{36 \times 10^{-3}\,[H^+]}{3.162 \times 10^{-7} + [H^+]} + \frac{14 \times 10^{-3}\,[H^+]}{5.011 \times 10^{-8} + [H^+]} + \frac{[Pi]\,[H^+]}{1.995 \times 10^{-7} + [H^+]}\right\} \quad (7)$$

The concentrations of the fixed buffers in frog muscle and all of the equilibrium constants used in this calculation are those given by Curtin and Woledge (10); $[H^+]$ and [Pi] are measured as explained above. We have neglected the CO_2/bicarbonate system, since it was shown by Stella (11) that if the surrounding medium does not contain CO_2 the intracellular [bicarbonate] is only about 1 mM. Note that Equations (6) and (7) take full account of the effects of PCr hydrolysis, because neither PCr nor Cr has a pK near to pH = 7. The increase in the third term of equations (6) and (7) as Pi is liberated from PCr shows how much extra lactate production was concealed by the "alkalinising" effect of PCr splitting. Note that Equations (5) - (7) do not involve any approximations or assumptions that restrict accuracy unless only small changes in $[H^+]$ have occurred. For this reason we consider our method of calculation to be simpler and more accurate than that of Piiper (see later in this volume), who has attacked the same problem in terms of the "buffering capacity" $\frac{d\,[BH^+]}{d\,(pH)} = \beta$, which he assumes to be constant over the range of experimental conditions encountered.

A detailed discussion of the two theoretical approaches will be given in the Discussion section following Professor Piiper's paper. Incidentally, it has now become possible to estimate LA production directly by proton NMR, (12) rather than indirectly from the change of pH interacting with the intracellular buffers present.

Total LA^- produced and total phosphorus utilised (P_{util}; the amount of ATP breakdown that would have been required in the absence of all

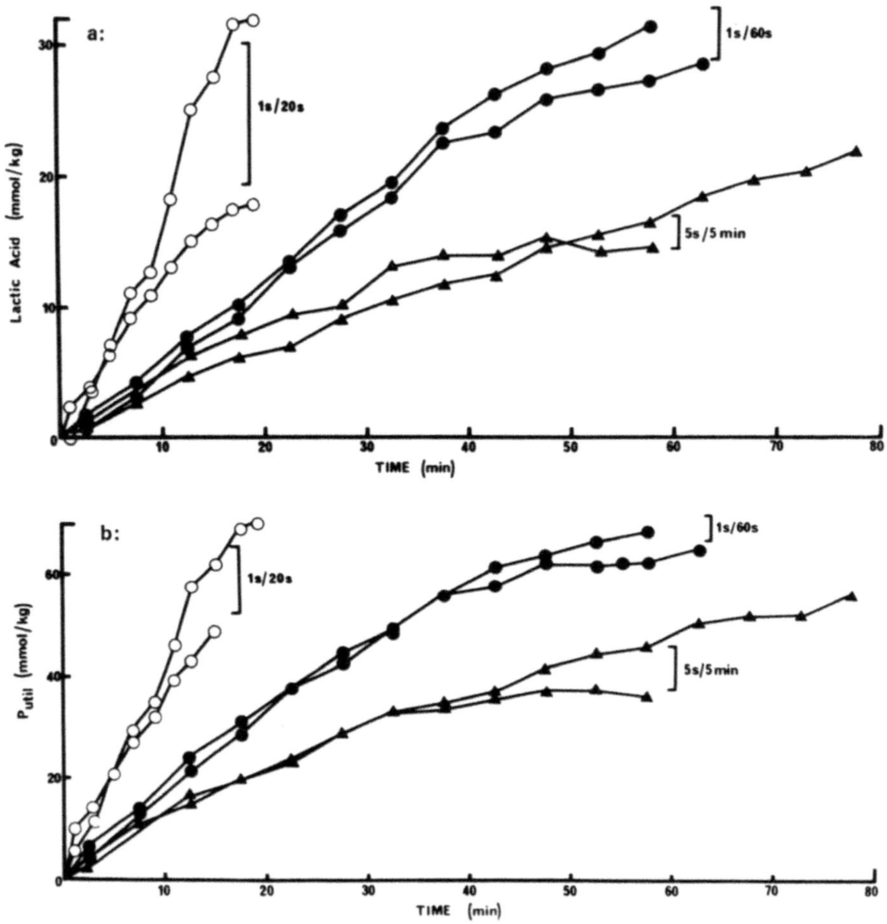

Fig. 5 a and b. Total lactic acid produced (a) and total phosphorus utilised (b) as a function of time for each of the six experiments

recovery processes) are shown in Fig. 5 as a function of time for each of the six experiments. P_{util} was calculated according to the following equation:

$$\Delta P_{util} = 3/2\Delta[LA] - \Delta[PCr] - \Delta[ATP] \tag{8}$$

(There is some doubt as to the amount of ATP resynthesised per mole of LA produced (13); however, until this question is resolved we think it best to use the commonly accepted stoichiometry). Both LA production and phosphate utilisation depend heavily upon the pattern of stimulation. Even in cases in which the muscles have been stimulated for the same total amount of time per hour (5 s/5 min and 1 s/60 s)

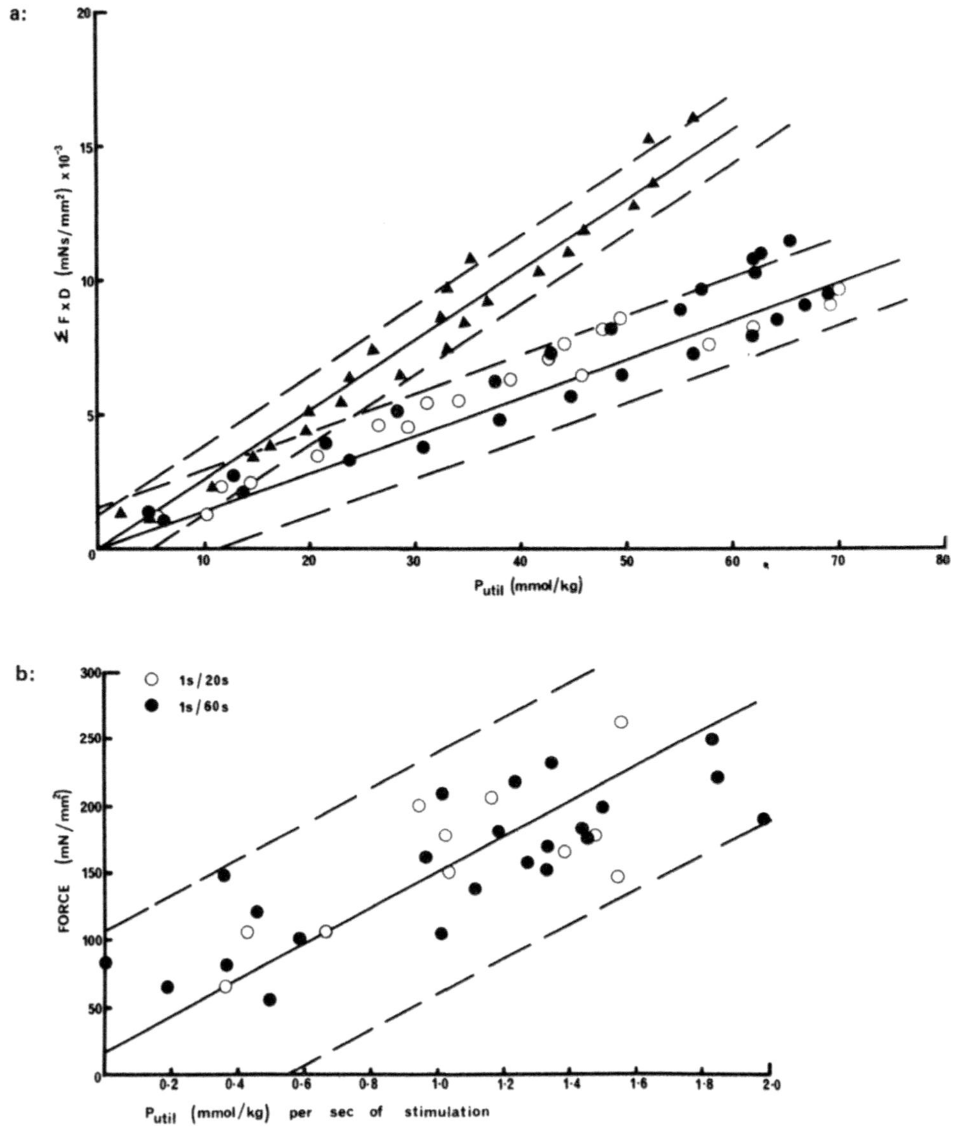

Fig. 6. a The sum of the force developed in each contraction x the duration of stimulation [Σ (FD)] as a function of phosphorus utilised (P_{util}). FD was obtained by multiplying the force developed at the midpoint of each stimulation by the duration of stimulation. This method was chosen both for its simplicity and because it does not assume that ATP is being hydrolysed during the relaxation period. ▲, 5 s/5 min; ●, 1 s/60 s; O, 1 s/20 s. The solid lines were obtained by least-square regression analysis, assuming that all the experimental error is associated with P_{util}. In both cases the intercepts were negligible. For 5 s stimulation Σ FD ($mN \cdot mm^{-2}$ s) = 261 P_{util} ($mM \cdot kg^{-1}$); r = 0.99; n = 25. For 1 s stimulation Σ FD ($mN \cdot mm^{-2}$ s) = 142 P_{util} ($mmol \cdot kg^{-1}$); r = 0.97; n = 49. The dashed lines are 95% confidence limits.

b Force as a function of P_{util} per second of stimulation, $\Delta P_{util}/D$, estimated directly from the slopes in Fig. 4 b. For clarity, only the date from stimulations of 1 s duration are included. For the 1 s/20 s experiments, the results there is a twofold difference in the total phosphorus utilised and in the total $\dot{LA}$ produced.

These results show that the factor(s) determining the rate of glycolysis under the conditions of our experiments are closely linked to the rate at which phosphorus is being utilised. In none of the patterns of stimulation, however, is the rate of glycolysis sufficient to protect the muscles from progressive loss of PCr, whose contribution to recovery becomes negligible by the time that force has declined to about 160 $mN \cdot mm^{-2}$. This change from two sources to a single source of rephosphorylation of ADP may cause the curvature in Figs. 3a and 4: the plots of PCr, Cr, Pi, ATP, and Affinity against force development all show a change in slope when force development declines to about 160 $mN\ mm^{-2}$. From the viewpoint of animal design it is surprising that glycolysis is not turned on more fully by the milder patterns of stimulation, for this would seem to provide a better strategem for conserving PCr to use in emergencies.

To relate our experimental results to theoretical schemes of contractility, we have examined the relation between force development and phosphorus utilisation in several different ways. In Fig. 6a, P_{util} is plotted against the cumulative sum of the force development in each contraction (F) times the duration of stimulation (D). There is a highly linear relation between these two parameters; the results fall into two groups according to whether D is 1 s or 5 s. For each duration of contraction the results can be represented by the equation:

$$P_{util} = \Sigma FD/k \tag{9}$$

where the slope k depends on the duration of contraction.

To relate force developed in a single contraction to the associated chemical change, it is merely necessary to consider the increments (FD and ΔP_{util}) in both axes of Fig. 6a that result from an additional contraction cycle:

---

were averaged over 4-min rather than the usual 2-min intervals. The solid line was obtained by least-squares regression analysis, assuming that all the experimental error is in P_{util}; slope = 134 $mN \cdot mm^{-2}$ mmol-1 kgs; r = 0.80; n = 34; $P \ll 0.001$. The dashed lines are 95% confidence limits and the intercept is not significant

$$FD = k\,\Delta P_{util} \text{ or } F = k\,\Delta P_{util}/D \quad (10)$$

The latter relationship is illustrated directly for 1 s contractions in Fig. 6b, where the values of ΔP_{util} have been taken from point-to-point gradients in Fig. 5b. The points closest to the origin in Fig. 6b represent the end rather than the beginning of the experiment in Fig. 6a. The direct interpretation of Equation 10 is thus that as the muscle becomes fatigued, the decline in force is proportional to the decline in rate of phosphorus utilisation, and the constant of proportionality (k) is considerably greater for the 5-s contractions ($k \cong 260$) than for the 1-s ones for which $k \cong 140$ (as determined from Fig. 6a or b).

The increased economy of contraction observed in the contractions of longer duration agrees qualitatively and quantitatively with evidence derived from studies of nonfatigued muscle. It has long been known that in single isometric tetani the rate of heat production declines after the first second of stimulation (14, 15); a similar pattern has been demonstrated for PCr hydrolysis (16, 17). Marechal's and Mommaerts' (16) results with single and repeated tetani of frog sartorii at $0^{o}C$ fitted the equation

$$\Delta PCr = 0.33 + 0.28D \text{ mmol} \cdot \text{kg}^{-1} \quad (11)$$

for values of D up to 60 s. In absolute terms, Equation 11 accords reasonably well with other measurements on frog sartorii by chemical methods at $0^{o}C$ (17, 18, 19) or by ^{31}P NMR at $4^{o}C$ (1).

Our present results for D = 1 s and 5 s can be analysed in a similar way:

$$P_{util} = F\,(4.01 + 3.03D) \times 10^{-3} \text{ mmol kg}^{-1} \quad (12)$$

This equation, together with the ratio between its constants, closely resembles Equation (11), and furnishes the additional information that both terms in the bracket are directly proportional to force development as it declines during fatigue. We cannot distinguish in the present experiments between phosphorus utilisation _during_ contraction and any that is utilised _between_ contractions, but we have reason from other experiments to think that the latter contribution is small. Studies of nonfatigued frog (ref. 16, p. 62) and toad (1) gastrocnemii at $4^{o}C$ support the view that these muscles break down 1.5-1.7 mmol PCr/kg during a 1-s contraction; this is in quantitative agreement with the present results and is roughly three times more than the values obtained on sartorii at 0°-$4^{o}C$.

In trying to relate our results to the fundamental processes thought to be occurring in contracting muscle, we found Equations 11 and 12 particularly attractive, since it seems natural to associate the first bracketed term with the chemical and mechanical events required to set up isometric force; while the second, duration-dependent term could be associated with the actual cross-bridge cycling that maintains the force.

Certainly there is good evidence that a quantity of PCr is hydrolysed at the onset of contraction even when actomyosin interaction has been prevented by stretching (20, 21). If the duration-dependent term in Equation (12) can indeed be equated with cross-bridge cycling, then our results show that the economy of maintenance of force remains constant in fatiguing muscle at about 300 Nmm^{-2} s per mol ATP/kg split, and that the force declines because the rate of ATP utilisation falls progressively. If so, anything that reduced the rate of ATP hydrolysis (e.g., product inhibition or decreased Ca^{+2} release) would automatically reduce the force proportionately and result in fatigue.

Conclusions

In this study we have used ^{31}P NMR together with other techniques to relate force development to biochemical and energetic changes in fatiguing frog muscle. We find that force development is closely correlated with metabolite levels rather than with any independent changes in excitatory conduction. We also find that the decline in isometric force development as the muscle fatigues is proportional to the decrease in the rate of phosphorus utilisation; this last result suggests that the hydrolysis of ATP in each cross-bridge cycle produces a fixed mechanical impulse (force x time) and that the economy of ATP hydrolysis remains unchanged. (A more expanded account of this work will appear in Nature.)

References

1. DAWSON, M.J., GADIAN, D.G., WILKIE, D.R.: Contraction and recovery of living muscles studied by ^{31}P nuclear magnetic resonance. J. Physiol. 267, 703-735 (1977)
2. HOULT, D.I., RICHARDS, R.E.: Critical factors in the design of sensitive high resolution nuclear magnetic resonance spectrometers. Proc. R. Soc. [A], 344, 311-340 (1975)
3. BURT, C.T., GLONEK, T., BARANY, M.: Analysis of phosphate metabolites, the intracellular pH and the state of adenosine triphosphate in intact muscle by phosphorus nuclear magnetic resonance. J. Biol. Chem. 251, 2584-2591 (1976)
4. BURT, C.T., GLONEK, T., BARANY, M.: ^{31}P nuclear magnetic resonance detection of unexpected phosphodiesters in muscle. Biochemistry (N.Y.) 15, 4850-4852 (1976)
5. VERGARA, J.L., RAPOPORT, S.I., NASSAR-GENTINA, V.: Fatigue and posttetanic potentiation in single muscle fibers of the frog. Am. J. Physiol. 232, C185-C190 (1977)
6. SPANDE, J.I., SCHOTTELIUS, B.A.: Chemical basis of fatigue in isolated mouse soleus muscle. Am. J. Physiol. 219, 1490-1495 (1970)
7. ALBERTY, R.A.: Standard Gibbs free energy, enthalpy and entropy changes as a function of pH and pMg for several reactions involving adenosine phosphates. J. Biol. Chem. 244, 3290-3302 (1969)
8. THE ROYAL SOCIETY: Quantities, Units and Symbols. London 1975
9. MAINWOOD, G.W., WORSLEY-BROWN, P., PATERSON, R.A.: The metabolic changes in frog sartorius muscles during recovery from fatigue at different external bicarbonate concentrations. Can. J. Physiol. Pharmacol. 50, 143-155 (1972)
10. CURTIN, N.A., WOLEDGE, R.C.: Energy changes and muscular contraction. Physiol. Rev. to be published
11. STELLA, G.: The combination of carbon dioxide with muscle: its heat of neutralization and its dissociation curve. J. Physiol. 68, 49-66 (1929/1930)
12. BROWN, F.F., CAMPBELL, I.D., KUCHEL, P.W., RABENSTEIN, D.C.: Human erythrocyte metabolism studies by ^{1}H spin echo NMR. FEBS Lett. 82, 12-16 (1977)
13. DE FURIA, R.R., KUSHMERICK, M.J.: ATP utilization associated with recovery metabolism in anaerobic frog muscle. Am. J. Physiol. 232, 1 (1977); Cell Physiol. 1, C30-C36 (1977)
14. HARTREE, W., HILL, A.V.: Regulation of the supply of energy in muscular contraction. J. Physiol. 55, 133-158 (1921)
15. AUBERT, X.: Le couplage energetique de la contraction musculaire. Brussels: Arscia 1956
16. MARECHAL, G., MOMMAERTS, W.F.H.M.: The metabolism of phosphocreatine during an isometric tetanus in the frog sartorius muscle. Biochim. Biophys. Acta 70, 53-67 (1963)
17. HOMSHER, E., RALL, J.A., WALLNER, A., RICCHIUTI, N.V.: Energy liberation and chemical change in frog skeletal muscle during single isometric contractions. J. Gen. Physiol. 65, 1-21 (1975)
18. GILBERT, C., KRETZSCHMAR, K.M., WILKIE, D.R., WOLEDGE, R.C.: Chemical change and energy output during muscular contraction. J. Physiol. 218, 163-193 (1971)
19. CURTIN, N.A., GILBERT, C., KRETZSCHMAR, K.M., WILKIE, D.R.: The effect of the performance of work on total energy output and metabolism during muscular contraction. J. Physiol. 238, 455-472 (1974)
20. HOMSHER, E., MOMMAERTS, W.F.H.M., RICCHIUTI, N.V., WALLNER, A.: Activation heat, activation metabolism and tension-related heat in frog semitendinosus muscles. J. Physiol. 220, 601-625 (1972)
21. SMITH, I.C.H.: Energetics of activation in frog and toad muscle. J. Physiol. 220, 583-599 (1972)
22. CHALOVICH, J.M., VURT, C.T., COHEN, S.M., GLONEK, T., BARANY, M.: Identification of an unknown ^{31}P nuclear magnetic resonance from dystrophic chicken as L-serine ethanolamine phosphodiester. Arch. Biochem. Biophys. 182, 683-689 (1977)

Chapter 3

Circulatory Responses to Exercise

J. Durand

This paper deals with two different aspects of the cardiovascular adjustments that take place during true and simulated exercise. The first is concerned with the cardiovascular effects of electric shock compared with those arising during normal exercise. The second tends to give a biological meaning to the slope and the intercept of the linear relationships linking cardiac output and oxygen consumption during the steady state of submaximal exercise.

Electric Shock

Some psychiatrists are again using electric shock as a possible treatment for depressive psychosis. Shocks are now administered to patients after curarisation, under a general anaesthetic and with assisted ventilation. In three patients, cardiac output (thermodilution), pulmonary and brachial arterial pressures, and heart rate (ECG) were measured before and as soon as possible after the shock. The results (Fig. 1) show that after the shock there is a sudden increase in cardiac output, followed by a prolonged decrease. Blood pressure changes follow the same general pattern. It is interesting to note that (1) there are no significant alterations in the peripheral resistance; (2) after a sudden rise, the heart rate falls below control values, whereas the calculated stroke volume increases and decreases more slowly.

It is generally accepted that cardiovascular adjustments at the start of exercise are under nervous control and do not depend upon the circulatory or metabolic changes taking place in the working muscles; however, there is still no general agreement as to whether these changes are initiated by proprioceptive afferences, directly by central motor activity, or by both (1).

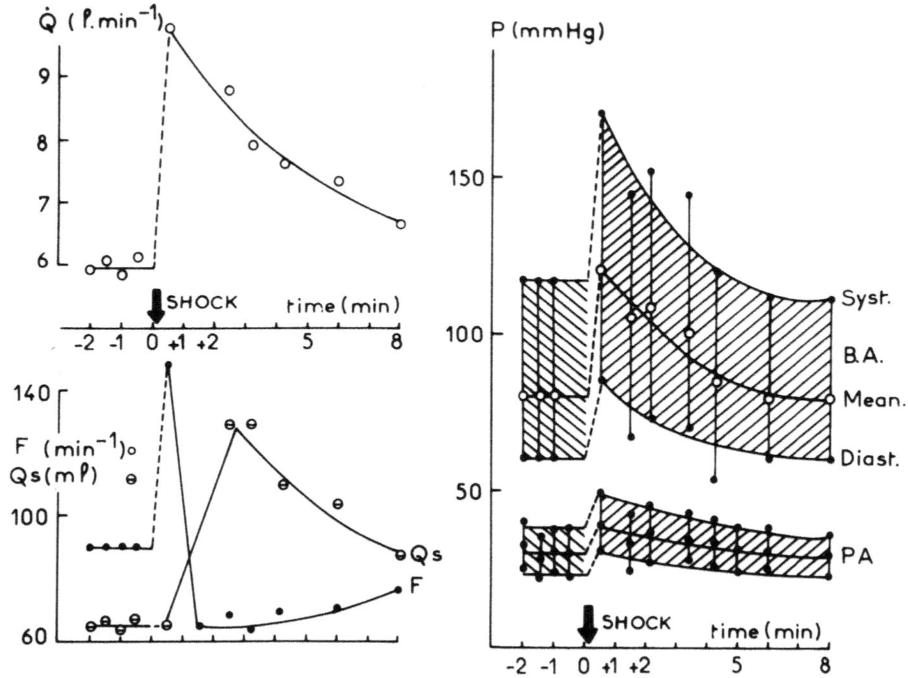

Fig. 1. Cardiovascular changes after therapeutic electric shock in an anaesthetised and curarised subject. $\dot{Q}$, cardiac output; F, heart rate; Q_S, calculated stroke volume; P, maximal (syst.) mean and minimum (diast.) blood pressure measured in the brachial (BA) and the pulmonary (PA) arteries

Of course, electric shock may not appear to be the most appropriate technique for selective activation of the motor structures of the cephalic pole of the central nervous system, and the data reported need further investigation before correct interpretation can be assured. However, a tentative explanation could be that cardiac responses at the onset of exercise are mainly, if not totally, controlled by the central nervous system; the sympathetic system is the main efferent, so that the increase in contractility matches the increase in heart rate, and the stroke volume does not change very much even when the heart rate increases suddenly (Fig. 2).

In the present study, an abrupt increase in cardiac output resulted in a sharp rise in blood pressure, which induced a reflex slowing of the heart; this in turn allowed a larger stroke volume to develop.

The absence of any change in peripheral resistance favours the assumption that blood flow redistribution is initiated mainly by propriosensitive afferences integrated at a suprasegmentary level. This

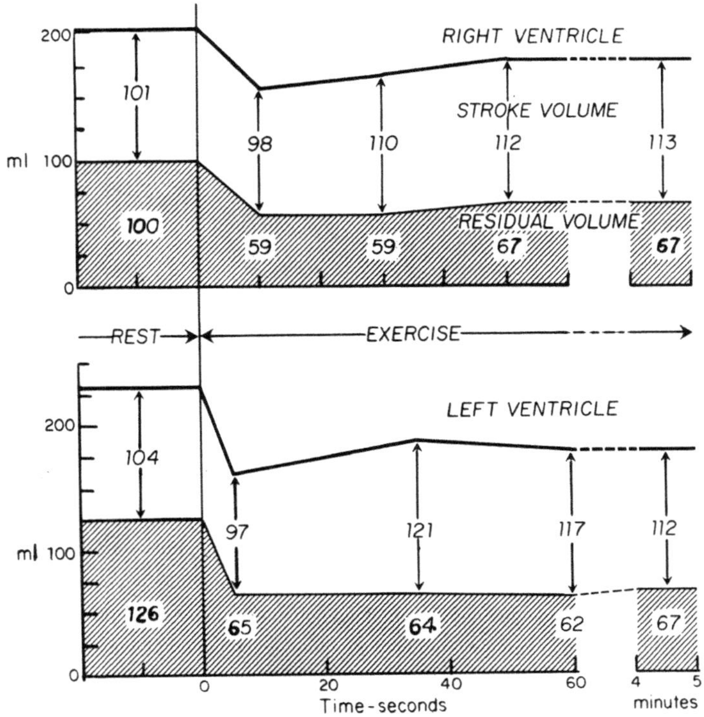

Fig. 2. Ventricular volumes in right (upper panel) and left (lower panel) ventricles at rest and at onset of exercise. Volumes were determined by radiocardiography (19). □, stroke volume; ▨, residual volume

hypothesis would agree with observations made in paraplegic subjects (2) and with results obtained in skin blood flow measurements in normal subjects at the start of exercise (3) (Fig. 3 and 4).

In summary, it seems that cardiac adjustments at the beginning of and probably also during exercise are centrally programmed according to the mumber of motor units that are, or will be, activated by the motor structures of the upper central nervous system, i.e., according to the strength of the muscular activity that will be or is being performed. It seems that the main efferent is the sympathetic system, and the adequacy of cardiac output and peripheral arteriolar flow is achieved by the barovagal loop.

Submaximal Exercise

In 1915 Lindhard established that during a steady state of muscular exercise there was a linear relationship between oxygen uptake ($\dot{V}_{O_2}$)

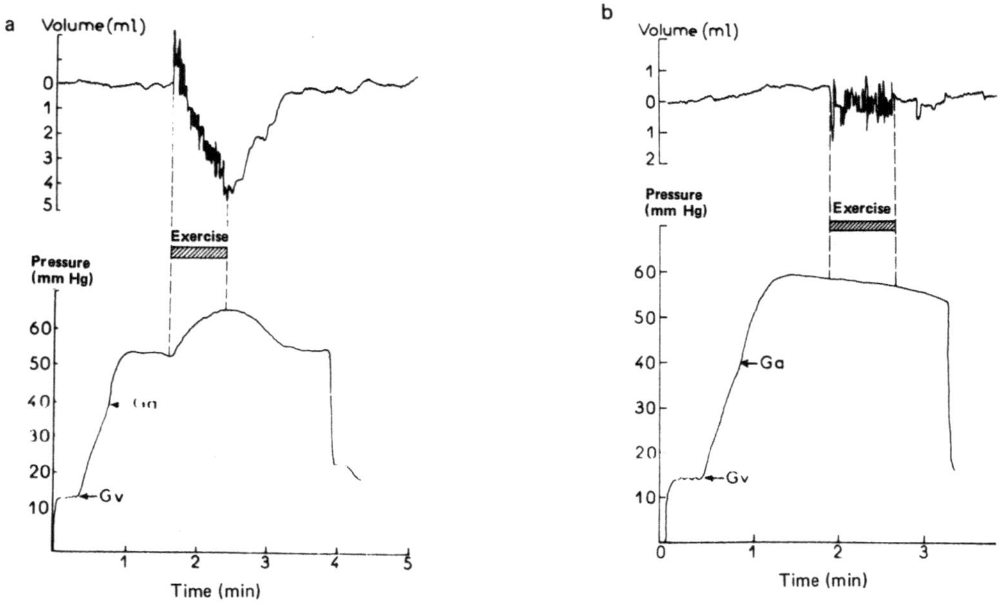

<u>Fig. 3.</u> <u>a</u> and <u>b</u> Changes in capacitance vessel tone during exercise in normal subjects (<u>a</u>) and in paraplegics (<u>b</u>).

Volume and venous pressure were measured in legs before, during and after an exercise with the upper limbs (▨).
(Volume, volume changes in millilitres, measured in the foot; pressure was measured in an isolated venous segment, isolated by inflation of two cuffs at times G_v and G_a.)

and cardiac output (Q) (4). Afterwards, many investigators confirmed this finding, using similar or different, and supposedly more accurate, techniques. This relationship may be written:

$$\dot{Q} = K\,\dot{V}_{O_2} + \dot{Q}_O \qquad (1)$$

K being the slope and $\dot{Q}_O$ the intercept with the ordinates.

On the other hand, the equation describing the Fick principle can be rearranged as:

$$\dot{Q} = \dot{V}_{O_2}/Ca_{O_2} + \dot{Q}\ C\bar{v}_{O_2}/Ca_{O_2} \qquad (2)$$

(Ca_{O_2} and $C\bar{v}_{O_2}$ are oxygen content in arterial and mixed venous blood, respectively).

The difference is analogous to that between the O_2 clearance in the lung and the O_2 "nonclearance" in the tissues.

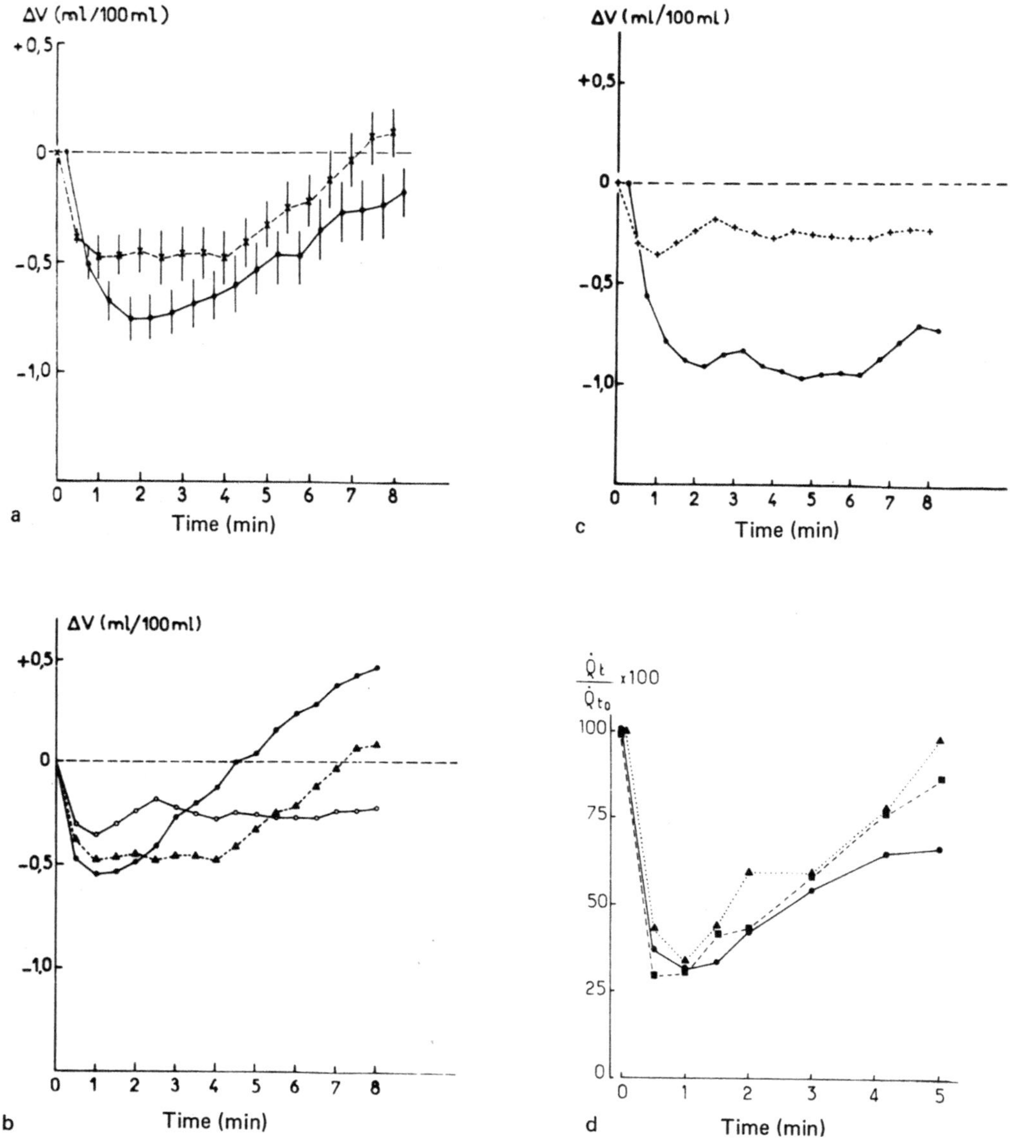

Fig. 4. a Volume changes (ΔV in millilitres per 100 millilitres of tissue) in the right hand at the onset of bicycling at the same force (30 W) but at different speeds, viz. 40 (x) and 60 (·) rpm). b Volume changes (ΔV in millilitres per 100 millilitres of tissue) in the right hand at the onset of bicycling at different forces, viz. 0 W, i.e., passive mobilisation of the legs (o-o) and 30 (▲-▲) and 90 (●-●) W) but at the same speed (40 rpm). c Volume changes (ΔV: in millilitres per 100 millilitres of tissue) in the right hand during passive mobilisation of the legs at 40 rpm (x) and 60 rpm (●), d Percentage change in skin (hand) blood flow ($\dot{Q}_t/\dot{Q}_{to}$) at onset of exercise at different forces, 100% being the control value in each case. The work load is expressed as a fraction of the maximal aerobic capacity of the subject (3). ● ●, O.HO $\dot{V}_{O_1}$; ■---■, 0.58 $\dot{V}_{O2}$; ▲···▲, 0.70 $\dot{V}_{O_2}$. All three together give $\dot{W}_{EM}$

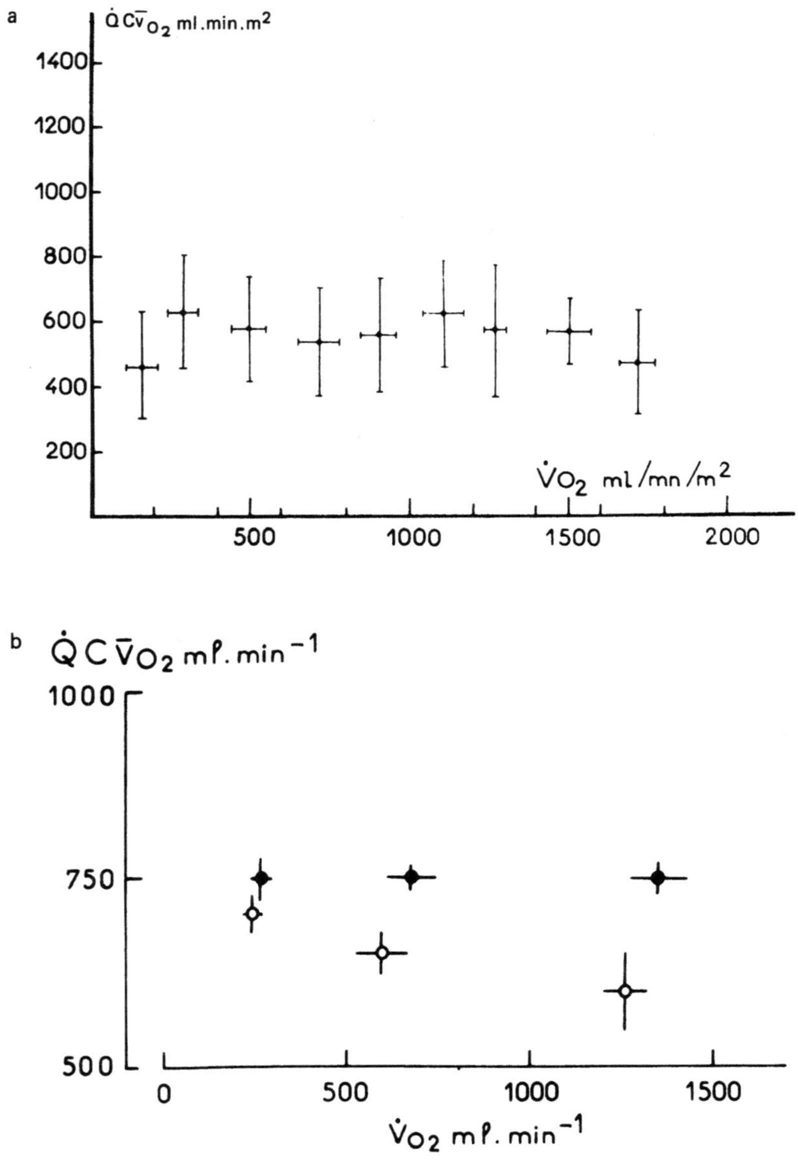

Fig. 5. a Flow of oxygen returning to the lung as a function of oxygen uptake, ($\dot{V}_{O_2}$: ml. min^{-1} m^{-2} body surface area) at rest and at different levels of muscular exercise. Oxygen flow is computed as the product of the cardiac output ($\dot{Q}$: ml. min.$^{-1}$ m^{-2}) and the concentration of oxygen in mixed venous blood (pulmonary artery). Means and standard deviations. b Flow of oxygen returning to the lung ($\dot{Q}$ $C\bar{v}_{O_2}$ ml. min^{-1}) as a function of oxygen consumption ($\dot{V}_{O_2}$ ml. min^{-1}). Measurements were taken at rest and at two levels of exercise in ten subjects before (•) and after (o) β-blockade. Means and standard error

On comparison of (1) and (2), $\dot{Q}\ C\bar{v}_{O_2}/Ca_{O_2}$ appears to be either a constant or a linear function of $\dot{V}_{O_2}$.

Data available in the literature and laboratory measurements appear to indicate that the first of these possibilities has been experimentally verified. Therefore, according to Equation (2), the slope, K, in relation (1), is the reciprocal of Ca_{O_2} and the intercept, $\dot{Q}_O$, is equal to $(Ca_{O_2})^{-1}\ \dot{Q}\ C\bar{v}_{O_2}$; this means that this product is constant, and independent of $\dot{V}_{O_2}$. As Ca_{O_2} does not change or changes only slightly during exercise, $\dot{Q}\ C\bar{v}_{O_2}$, i.e., the amount of oxygen flowing back to the lung in venous blood per unit of time, is presumably also constant and independent of $\dot{V}_{O_2}$ (Fig. 5).

This fits in well with the values determined empirically for the same parameters; for instance, in normal subjects of both sexes exercising in the supine position, DONALD et al. (5) found that cardiac output was related to oxygen uptake by the following regression equation:

$$\dot{Q} = 5.34\ \dot{V}_{O_2} + 3.71$$

Use of the same data and $(Ca_{O_2})^{-1}$ as slope and $(Ca_{O_2})^{-1}\ \dot{Q}\ C\bar{v}_{O_2}$ as intercept would give the following relation:

$$\dot{Q} = 5.30\ \dot{V}_{O_2} + 3.62$$

Using the studies by BARGER et al. and LEUSEN et al. on unanaesthetised dogs running on a treadmill, DONALD et al. (5) derived a very similar relation:

$$\dot{Q} = 5.38\ \dot{V}_{O_2} + 3.04$$

Thus this relation does not seem to be specific for the human species (6).

Physiological or pathological alterations of Ca_{O_2} result in corresponding changes in the slope (e.g., 5, 7, 8, 9, 10, 11, 12).

For subjects exercising in a supine position ($\dot{Q}$ is expressed in $ml.mn^{-1}m^{-2}$ of body surface area):

$$\dot{Q}_{sup} = 5.48\ \dot{V}_{O_2} + 3.15 \qquad Ca_{O_2} = 0.182 \pm 0.007$$

While during exercise in a sitting position

$$\dot{Q}_{si++} = 5.29\ \dot{V}_{O_2} + 2.75 \qquad Ca_{O_2} = 0.189 \pm 0.005$$

and during exercise in a standing position

$$\dot{Q}_{s+} = 5.18\ \dot{V}_{O_2} + 2.41 \qquad Ca_{O_2} = 0.192 \pm 0.005$$

The relation is slightly different for men (M) and women (F) in a sitting position:

$$\dot{Q}_{F} = 5.95\ \dot{V}_{O_2} + 2.44 \qquad Ca_{O_2} = 0.171 \pm 0.005$$

$$\dot{Q}_{M} = 5.32\ \dot{V}_{O_2} + 2.73 \qquad Ca_{O_2} = 0.190 \pm 0.003$$

There are exceptions, e.g., when cardiac output is increased by emotional stress or heat exposure. In cardiac failure the $\dot{Q}$, $\dot{V}_{O_2}$ relationship is no longer linear, and $\dot{Q}\ C\bar{v}_{O_2}$ falls during exercise; this condition is mimicked by β-blocking agents (Fig. 5b).

Natural or simulated altitude exposure induces notable alterations in $\dot{Q}\ C\bar{v}_{O_2}$ (Fig. 6) (13, 14, 15, 16, 17, 18). During acute exposure $\dot{Q}\ C\bar{v}_{O_2}$ is still independent of $\dot{V}_{O_2}$, but at a lower level than at sea level. In people resident at high altitudes, in contrast, $\dot{Q}\ C\bar{v}_{O_2}$ values are higher, as polycythaemia overcompensates for decreases in oxygen arterial saturation, and when an hyperoxic mixture is inspired, to give an inspired partial pressure of oxygen equal to that at sea level, $\dot{Q}\ C\bar{v}_{O_2}$ rapidly exceeds the control value. This is because the cardiac output does not decrease or decreases only slightly. This observation favours the assumption that metabolic demand or oxygen content in venous blood plays little or no role in cardiac control during exercise.

In conclusion, in normal subjects cardiac output is very probably under nervous control both during muscular activity and at the onset of exercise. Cardiac and skeletal muscles are controlled in parallel, so that cardiac output and arteriovenous difference for oxygen are equally involved in increasing oxygen uptake and supply.

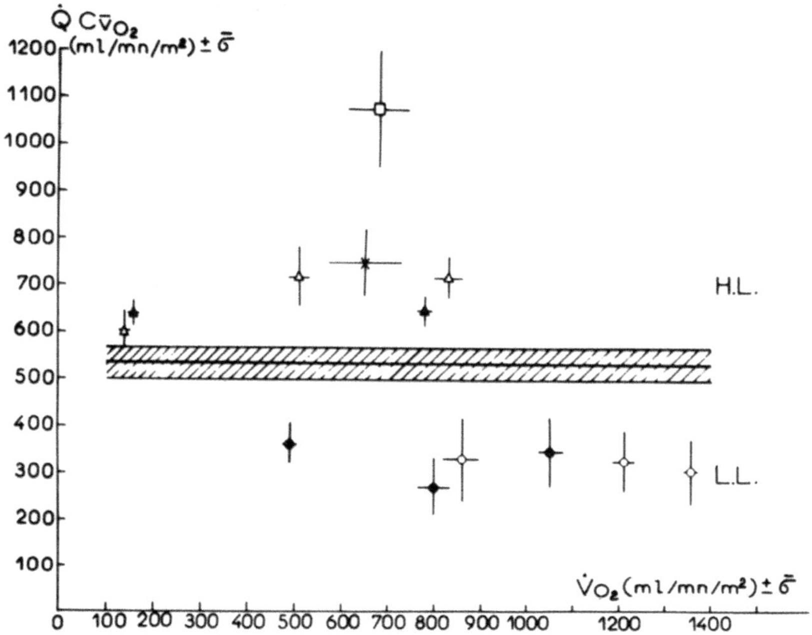

Fig. 6. Flow of oxygen returning to the lung ($\dot{Q}\ C\bar{v}_{O_2}$ 1 ml. $min^{-1}\ m^{-2}$) in subjects exposed to hypobaric conditions. Horizontal line and hatched area indicate value obtained at sea level (mean and standard error). Lowlanders (LL) were acutely exposed to altitude (3800 m; •) or performed exercises in an hypobaric chamber (4000 m; o). Highlanders (HL) were examined at rest or during exercise at 3800 m (▲) or 4300 m (△). Values were obtained in HL at 3800 m before correction of hypoxia (x) and after correction by breathing a hyperoxic mixture (□). (Mean and standard error.)

Summary

Circulatory changes induced by electric shock suggest an initial central control of cardiac responses at the onset of exercise, whereas peripheral vasomotor changes are more probably controlled by peripheral afferences.

The flow of oxygen returning to the lung (cardiac output multiplied by oxygen concentration in mixed venous blood) during the steady state of submaximal exercise is constant, i.e., independent of the oxygen uptake. The slope of the relationship between cardiac output and oxygen consumption is equal to the reciprocal of the oxygen concentration in arterial blood.

References

1. Symposium on the "Onset of exercise". Délégation générale à la recherche scientifique et technique (eds). Toulouse 1971 (197 pages)
2. BIDART, Y., MAURY, M.: The circulatory behaviour in complete chronic paraplegia. Paraplegia 11, 1-24 (1973)
3. VALLADARES, V.R.W.: Débit sanguin de la main au cours de l'exercice musculaire à la température ambiante de 24°C. Diplôme d'Etudes et de Recherches en Biologie Humaine, U.E.R. Médicale Paris-Sud - Paris 1977 (70 pages)
4. LINHARD, J.: Über das Minutenvolumen des Herzens bei Ruhe und bei Muskelarbeit. Arch. Ges. Physiol. 161, 233-383 (1915)
5. DONALD, K.W., BISHOP, J.M., CUMMING, G., WADE, O.L.: The effect of exercise on the cardiac output and circulatory dynamics of normal subjects. Clin. Sci. 14, 37-73 (1955)
6. WADE, O.L., BISHOP, J.M.: The cardiac output during exercise. In: Cardiac output and regional blood flow. Blackwell Scientific Publication Oxford (eds), pp. 51-71 (1962)
7. ASTRAND, P.O., CUDDY, T.E., SALTIN, B., STENBERG, J.: Cardiac output during submaximal and maximal work. J. Appl. Physiol. 19, 268-274 (1964)
8. CHAPMAN, C.B., FISHER, J.N., SPROULE, R.J.: Behaviour of stroke volume at rest and during exercise in human beings. J. Clin. Invest. 39, 1208-1213 (1960)
9. DAMATO, A.N., GALANTE, J.G., SMITH, W.M.: Hemodynamic response to treadmill exercise in normal subjects. J. Appl. Physiol. 21, 959-966 (1966)
10. HOLMGREN, A.: Cardiorespiratory determinants of cardiovascular fitness. Can. Med. Assoc. J. 96, 697-702 (1967)
11. REEVES, J.T., GROVER, R.F., FILLEY, G.F., BLOUNT, S.G.: Circulatory changes in man during mild supine exercise. J. Appl. Physiol. 16, 279-282 (1961)
12. WANG, Y., MARSHALL, R.J., SHEPPERD, J.T.: The effects of changes in posture and grades exercises on stroke volume in man. J. Clin. Invest. 39, 1051-1061 (1960)
13. ASMUSSEN, E., NIELSEN, M.: The cardiac output in rest and work at low and high oxygen pressures. Acta Physiol.. Scand. 35, 73-83 (1955)
14. BANCHERO, N., SIME, F., PENALOZA, D., CRUZ, J., GAMBOA, R., MORTICORENA, E.: Pulmonary pressure, cardiac output and arterial oxygen saturation during exercise at high altitude and at sea level. Circulation 33, 249-262 (1965)
15. MORET, P., COVARRUBIAS, E., COUDERT, J., DUCHOSAL, F.: Cardiocirculatory adaptation to chronic hypoxia. Acta Cardiol. (Bruxelles) 27, 283-596 (1972)
16. PUGH, L.G.C.E.: Cardiac output in muscular exercise at 5800 m (19000 feet). J. Appl. Physiol. 19, 441-447 (1964)
17. STENBERG, J., EKBLOM, B., MESSIN, R.: Hemodynamic response to work at simulated altitude (4000 m). J. Appl. Physiol. 21, 1589-1594 (1966)
18. VOGEL, J.H.K., WEAVER, W.F., ROSE, R.L., BLOUNT, S.G., GROVER, R.F.: Pulmonary hypertension on excretion in normal living at 10050 feet (Leadville Colorado). Med. Thorac. 19, 461-477 (1962)
19. DONATO, L., PARKER, J.O., DURAND, J., ROCHESTER, D.F., COURNAND, A.: Tentative de mesure séparée des volumes de sang présents dans les deux ventricules pendant les différentes phases du cycle cardiaque, à l'etat stable et au cours de l'exercice musculaire modéré. Arch. Sci. Physiol. 15, 55-78 (1961)

Chapter 4

Production of Lactic Acid in Heavy Exercise and Acid-Base Balance

J. Piiper

Stoichiometric Relationships in Production and Metabolic Removal of Lactic Acid

It is important to realise that the substance produced in anaerobic glycolysis is, as far as stoichiometry is concerned, lactic acid, which dissociates into equimolar amounts of H^+ and lactate:

$$\left.\begin{array}{l} 1/2\ C_6H_{10}O_5 + 1/2\ H_2O \\ \text{glycogen monomer} \\ \\ 1/2\ C_6H_{12}O_6 \\ \text{glucose} \end{array}\right] \rightarrow \underset{\text{lactic acid}}{CH_3CHOHCOOH} \rightarrow \underset{\text{lactate}}{CH_3CHOHCOO^-} + H^+$$

Therefore, the generation of lactate must be accompanied by equimolar generation of H^+ ions, producing metabolic acidosis. The same is true for metabolic removal of lactate by resynthesis into glycogen (a) or by complete oxidation (b):

(a) $H^+ + CH_3CHOHCOO^- \rightarrow 1/2\ C_6H_{10}O_5 + 1/2\ H_2O$

(b) $H^+ + CH_3CHOH\ COO^- + 3\ O_2 \rightarrow 3\ CO_2 + 3\ H_2O$

Thereby the metabolic acidosis produced by anaerobic glycolysis is diminished or abolished.

It follows that the addition of lactate to a metabolising system must ultimately lead to metabolic alkalosis, since to each mole of lactate transformed one mole of OH^- (or HCO_3^-) is formed. Therapeutic use is made of this relationship when sodium lactate is infused for the correction of acidosis. On the other hand, this forced coupling of lactate and H^+ metabolism makes studies of metabolic effects of lactate per se, as distinguished from those of pH, rather difficult, at least in closed systems.

Quantitative Analysis of Buffering

The buffering power is characterised by the buffer value, β, which is defined as the number of H^+ ions bound per volume of buffer solution (or water), $\Delta[H^+]_b$, and per change of pH. Since practically all added H^+ is bound to buffers, the buffer value may also be expressed as the concentration change of the basic (A^-) or acidic form (HA) of the buffer system ($H^+ + A^- \rightleftharpoons HA$) per change of pH:

$$\beta = [H^+]_b/(-\Delta pH) = \Delta[A^-]/\Delta pH = \Delta[HA]/(-\Delta pH)$$

In analysis of physiological buffering it is appropriate to distinguish between the buffer system bicarbonate-carbon dioxide (reaction: $H^+ + HCO_3^- \rightleftharpoons CO_2 + H_2O$) and all other buffer systems (mostly proteins, inorganic and organic phosphates), generically designated as nonbicarbonate buffers (reaction: $H^+ + B^- \rightleftharpoons HB$).

At constant P_{CO_2} the buffer value for bicarbonate buffering, β_{Bic}, is easily calculated from the initial bicarbonate concentration, $[HCO_3^-]$, and the pH change, ΔpH, according to the following relationship derivable from the Henderson-Hasselbalch equation:

$$\beta_{Bic} = -\Delta[HCO_3^-]/(-\Delta pH) = [HCO_3^-] \cdot (10^{\Delta pH} - 1)/\Delta pH$$

The nonbicarbonate buffer value, β_{NB}, in most cases called simply "buffer value", is defined as

$$\beta_{NB} = -\Delta[B^-]/(-\Delta pH)$$

In contrast to β_{Bic}, β_{NB} has be determined for the tissue or compartment involved, by titration with fixed acid or with CO_2. In many cases β_{NB} may be considered as constant within a limited pH range.

The total buffer value, β_{tot}, is obtained by addition,

$$\beta_{tot} = \beta_{Bic} + \beta_{NB}$$

KARLSSON (1) reported that in human quadriceps muscle biopsies taken after heavy exercise the maximum lactate concentration was about 20 mEq/kg muscle, which corresponds to 33 mEq/kg intracellular muscle water. To estimate the acidification produced by the equivalent amount of H^+ assumed to remain entirely inside the muscle fibre, the nonbicarbonate buffer value of the intracellular muscle water determined in rat diaphragm muscle by HEISLER and PIIPER (2), 67 mEq/ (pH · kg H_2O), is used.

The calculated decrease in pH, visualised in Fig. 1, is 0.43 pH units, meaning a pH change from 6.90 (typical value for intracellular pH of resting muscle) to 6.47. These calculations are quite consistent with measurements of pH in human muscle biopsy taken after maximal exercise: the pH fell from 6.93 (control value) to about 6.40 (3).

It is evident from Fig. 1 that the role of bicarbonate buffering in the intracellular compartment of muscle is minor. This is due to the low bicarbonate concentration compared with that in the extracellular space and in the blood.

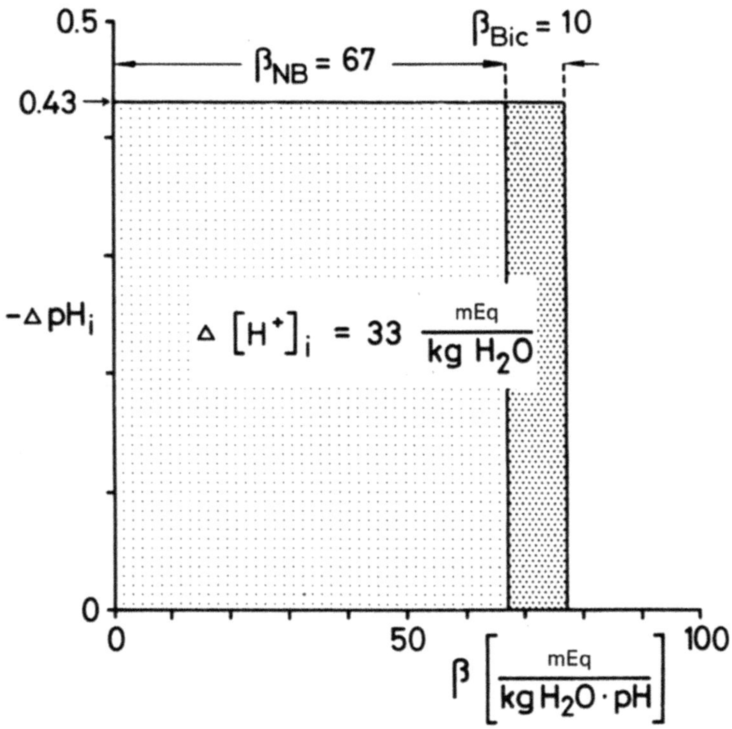

Fig. 1. Estimation of fall in pH (-ΔpH) by lactic acid, produced in human leg muscle by heavy exercise. For explanations, see text. Areas designate H^+ bound per kilogram intracellular muscle water, to nonbicarbonate buffers (▫) (lightly stippled) and to bicarbonate buffer (▩)

Effect of the Hydrolysis of Phosphocreatine on Buffering

The maximum acidification calculated above is expected to be modified by other chemical reactions occurring in exercising muscle. The hydrolysis of phosphocreatine is presumably of particular importance, because it generates OH^- ions according to the equation:

$$1 \text{ Phosphocreatine} \rightleftharpoons 1 \text{ Creatine} + 1 \text{ Phosphate} + x\ OH^-$$

According to MEYERHOF and LOHMANN (4) and LIPMANN and MEYERHOF (5), the value of x, i.e., the amount of OH^- generated per amount of phosphocreatine hydrolysed, is 0.4 at pH 6.8, and 0.6 at pH 6.4.

In heavily exercising muscles a large proportion of the phosphocreatine is hydrolysed. According to the human muscle biopsy data of KARLSSON (1), when lactate was increased by 20 mEq, creatine phosphate decreased by 16 mEq/kg, corresponding to 26 mmol/kg intracellular muscle water. Its effect on the change in pH is shown in Fig. 2. Hydrolysis of creatine phosphate shifts the buffer line to

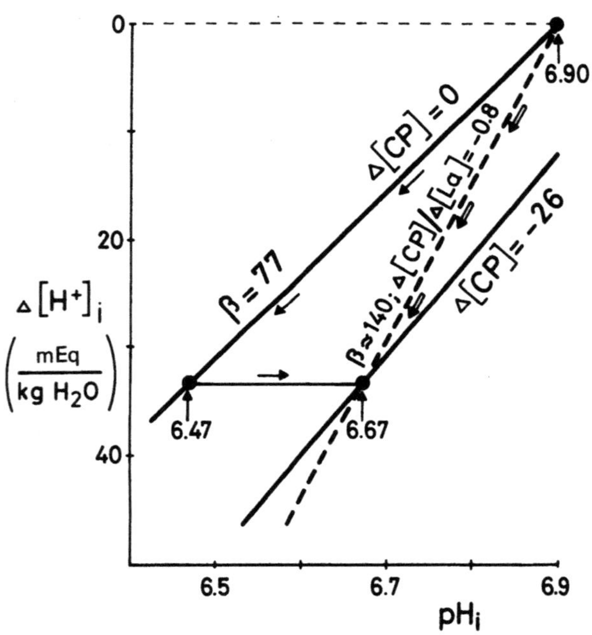

Fig. 2. Estimation of the effect of phosphocreatine hydrolysis upon buffering of H^+ ions derived from lactic acid formation. Abscissa, intracellular pH. Ordinate, buffered H^+ per kg intracellular water. Heavy lines, buffer lines for resting muscle (Δ[CP] = 0) and for exhausted, phosphocreatine-depleted muscle (Δ[CP] = -26). Broken line, "apparent buffer line" for transition from rest to exercise. β values in mEq/(pH·kg intracellular water). For further explanations, see text

the right, whereby H^+ can be bound at constant pH (according to the vertical distance of the buffer lines), or the fall in pH due to the addition of H^+ is reduced (according to the horizontal distance of the buffer lines). In the example, the overall fall in pH is reduced by 46%, the pH being restored from 6.47 to 6.67.
The relatively rapid resynthesis of phosphocreatine after the end of exercise (6, 7) can be expected to aggravate lactic acid acidosis. Conversely, the decrease in creatine phosphate at the beginning of exercise is expected to produce intracellular alkalosis, which may be viewed as a preparation of the muscle for subsequent lactic acid acidosis. Other reactions may be involved besides the hydrolysis of phosphocreatine, but presumably others are of less importance. Thus formation of free ammonia could bind H^+ ions, whereas hydrolysis of ATP to ADP and inorganic phosphate generates H^+ ions.

Outflow of H^+ from Muscle Fibres

If added (or generated) H^+ ions are distributed in a multicompartment system the pH change of each compartment (of volume V) results from the distribution of added H^+ and the appropriate buffer value:

$$H^+_b = V_1 \cdot \beta_1 \cdot (-\Delta pH_1) + V_2 \cdot \beta_2 \cdot (-\Delta pH_2) + \ldots$$

On the assumption that all generated H^+ ions remain inside the muscle, but are distributed also to the extracellular space, the intracellular fall in pH has been estimated to be reduced from 0.43 units to 0.38 units. A further reduction of the fall in pH is achieved when H^+ is allowed to escape into blood, to be buffered by blood buffer systems (8).

Before meaningful calculations are possible, more information is required than is available at present.

1) The β_{NB} values must be known. For accurate assessment of β_{Bic}, changes of P_{CO_2} should be known.
2) The distribution of H^+ to compartments will remain unknown until knowledge of the equilibrium partition of H^+ is achieved. Variable patterns of interrelationship between intracellular and extracellular pH have been reported. In most cases the ratio $\Delta pH_i : \Delta pH_e$

in skeletal muscle is about 0.5, but it may be close to zero in certain conditions (9, 10).

3) Equilibrium distribution of H^+ is probably never achieved, since during the equilibration metabolic processes involving lactate and H^+ will be initiated. Therefore, what is required for an analysis is the time course of pH (and eventually of β) in the main body compartments and organs involved.

Transfer Patterns of H^+ and Lactate

The transfer of H^+ and lactate between organ and body compartments is not dependent on the condition of equimolarity, which results from reaction stoichiometry (electroneutrality is easily achieved by the movement of other ions). Indeed, experiments have revealed large discrepancies in behaviour between buffered H^+ ions and lactate ions.

Thus in simultaneous measurements of efflux of H^+ and lactate from stimulated isolated rat diaphragm muscle, the efflux of H^+ was found to be faster than that of lactate by more than 1 order of magnitude (II). Another striking example of difference in the time course of changes in lactate and bound H^+ (= base deficit) was found in arterial blood of the elasmobranch fish (Scyliorhinus stellaris) after stimulation to exhaustion (Fig. 3) (12). Both $\Delta[H^+]$ and lactate began to increase at about the same rate, but $\Delta[H^+]$ levelled off in the first hour and then decreased, whereas lactate was still increasing. After 4-8 h there was practically no base deficit ($\Delta[H^+] \approx 0$), but lactate was at its maximum, amounting to 10-20 mEq/litre. The fate of the H^+ ions (buffering in intracellular space, excretion into water through gills or kidneys) could not be definitively clarified.

These and other findings make it desirable to examine the relationships between base deficit and lactate concentration in man and in mammals in more detail, particularly in unsteady states with rapidly changing concentrations.

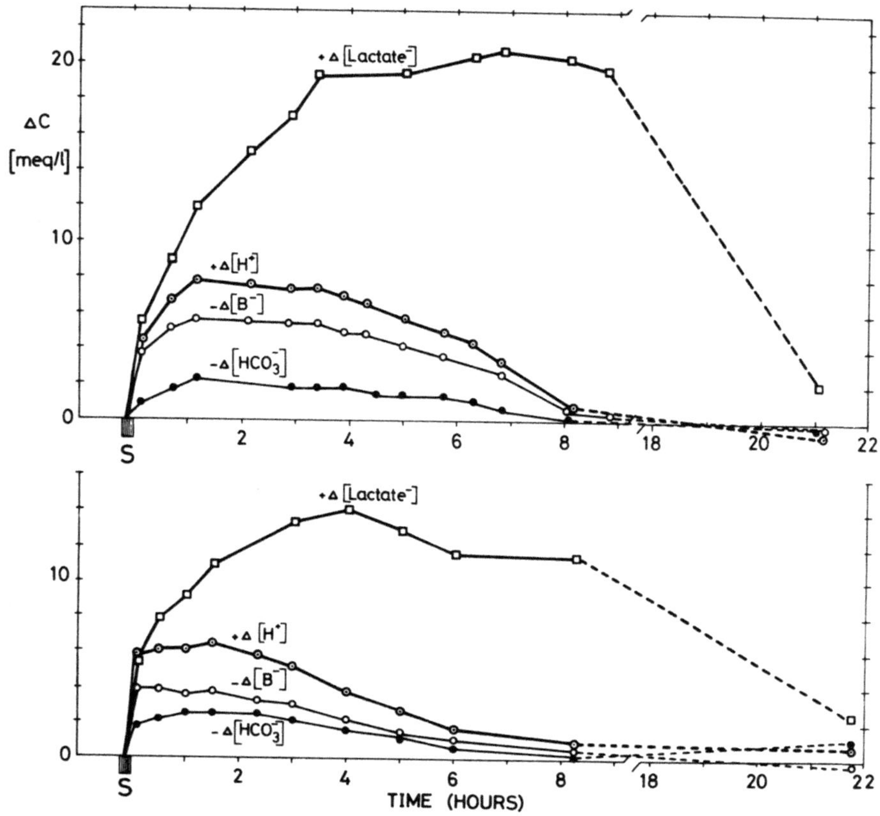

Fig. 3. Time course of increase of lactate (+ Δ[Lactate$^-$]), of total buffered H^+ (= base deficit) (+ Δ[H^+]) and of its components, H^+ buffered by nonbicarbonate buffers (- Δ[B^-]) and by bicarbonate buffers (-Δ[HCO_3^-]) in arterial blood of two dogfish *Scyliorhinus stellaris* stimulated (S) to exhaustion. See text. (J. Piiper et al., Respir. Physiol. 16, 290-303 (1972)

Summary

The production of lactic acid by anaerobic glycolysis in muscle fibres leads to the generation of equimolar amounts of H^+ ions and lactate. Similarly, when lactate is removed by complete oxidation or by resynthesis of glycogen, equivalent amounts of H^+ will disappear.

The H^+ ions are buffered by the nonbicarbonate and bicarbonate buffers of the muscle fibres, the resulting decrease in pH being determined by the buffer value.

The hydrolysis of phosphocreatine occurring at the onset of heavy exercise is expected to decrease the acidification by formation of OH^- ions.

Transfer of H^+ from the muscle fibres into the extracellular space of muscle, into blood, and into extracellular and intracellular spaces of other organs will progressively reduce the acidification of muscle fibres.

There is evidence for considerable differences in equilibration kinetics and transfer pathways between H^+ ions and lactate ions; therefore, a base deficit in blood may deviate from the equivalent lactate concentration.

References

1. KARLSSON, J.: Lactate and phosphagen concentrations in working muscle of man. Acta Physiol. Scand. [Suppl.] 358, 1-72 (1971)
2. HEISLER, N.; PIIPER, J.: The buffer value of rat diaphragm muscle tissue determined by P_{CO_2} equilibration of homogenates. Respir. Physiol. 12, 169-178 (1971)
3. HERMANSEN, L.; OSNES, J.B.: Blood and muscle pH after maximal exercise in man. J. Appl. Physiol. 32, 304-308 (1972)
4. MEYERHOF, O.; LOHMANN, K.: Über die natürlichen Guanidinophosphorsäuren (Phosphagene II). Mitteilung. Die physikalisch-chemischen Eigenschaften der Guanidinophosphorsäuren. Biochem. Z. 196, 49-72 (1928)
5. LIPMANN, F.; MEYERHOF, O.: Über die Reaktionsänderung des tätigen Muskels. Biochem. Z. 227, 84-109 (1930)
6. HULTMAN, E.; BERGSTRÖM, J.; ANDERSON, N. McL.: Breakdown and resynthesis of phosphorylcreatine and adenosine triphosphate in connection with muscular work in man. J. Clin. Lab. Invest. 19, 56-66 (1967)
7. PIIPER, J.; SPILLER, P.: Repayment of O_2 dept and resynthesis of high-energy phosphates in gastrocnemius muscle of the dog. J. Appl. Physiol. 28, 657-662 (1970)
8. PIIPER, J.: Buffering of lactic acid produced in exercising muscle. In: Symposium on the onset of exercise. Gibert, A., Guille, P. (eds.), pp 175-185. Toulouse: 1972
9. HEISLER, N.; PIIPER, J.: Determination of intracellular buffering properties in rat diaphragm muscle. Am J. Physiol. 222, 747-753 (1972)
10. HEISLER, N.: Intracellular pH of isolated rat diaphragm muscle with metabolic and respiratory changes of extracellular pH. Respir. Physiol. 23, 243-255 (1975)
11. HEISLER, N.: Kinetics of the efflux of hydrogen and lactate ions from isolated rat diaphragms stimulated in anoxia. Pfluegers Arch. 339, 51 (1973)
12. PIIPER, J.; MEYER, M.; DREES, F.: Hydrogen ion balance in the elasmobranch scyliorhinus stellaris after exhausting activity. Respir. Physiol. 16, 290-303 (1972)

Discussion

Because of lack of time following Professor Piiper's presentation, this discussion was held in private. Even so it was very brief and an opportunity never arose to explain coherently what I was trying to suggest.

Both Professor Piiper and our NMR group (see Dawson, Gadian and Wilkie, pp. 10-24 of this volume) have been concerned with the self-same problem, which is the interaction of lactic acid (LA) formation and phosphocreatine (PCr) hydrolysis with the intramuscular buffers, and the consequent changes in internal pH. We have used our NMR measurements of internal pH and of the change in inorganic phosphate concentration ([Pi]) as a new technique for estimating LA production in vivo.

Naturally I agree with much of what Professor Piiper said. My criticism is that I consider the principle of his method of calculation to be basically, and quite unnecessarily, inaccurate. He has based it on the concept of "buffer value" (β) first introduced by van Slyke (1922), where, for a fixed (i.e., nonbicarbonate) buffer, such as histidine, (its dissociated form B happens to be uncharged)

$$BH^{+} \rightleftharpoons B + H^{+} \qquad (1)$$

Professor Piiper writes for its buffer value

$$\beta = \Delta[B]/\Delta pH \qquad (2)$$

Strictly speaking, Equation (2) should be

$$\beta = d[B]/d(pH) = -d[BH^{+}]/d(pH) \qquad (3)$$

a differential quantity (see van Slyke, 1922, p. 528), not only as a matter of principle, but because β varies considerably with the pH (see Fig. 1).

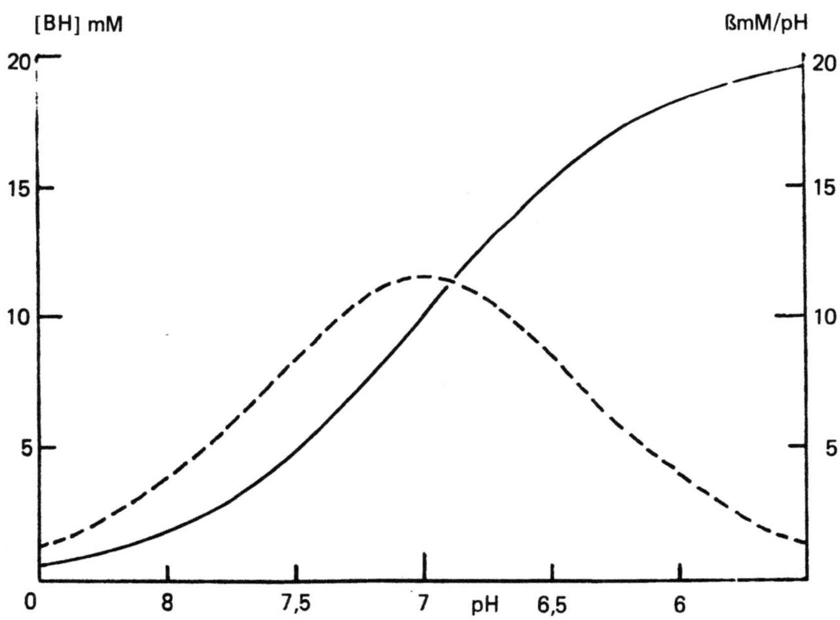

Fig. 1. The effect of adding acid to a buffer with a total concentration of 20 mM, expressed either in terms of [BH] (Eq. 4) (solid line) or of buffering power β (Eq. 7) (broken line)

It is the assumption made by Professor Piiper that β remains constant over a wide pH range (6.9-6.47) that is the basis for my criticism.

We have approached the same problem simply by going back to the basic equations (see our Equations (5)-(7).

For a single buffer we have

$$[BH^+] = C \times [H^+]/(K + [H^+]) \tag{4}$$

where the dissociation constant $K = B\ [H^+]/[HB^+]$ (5)

and the total analytical concentration of buffer $C = [B] + [BH^+]$ (6)

We think that our approach has the following advantages:

1) It is simple and makes no approximations
2) It makes no assumptions that limit its application to small variations in pH
3) It lends itself readily to the situation in which several different buffers are present, as is the case in muscle and elsewhere
4) It lends itself readily to the situation in which one or more of the buffers is changing in amount, as Pi does when PCr is hydrolysed.

To see the relationship between the two approaches, it is merely necessary to differentiate Equation 4:

$$-\frac{d[BH^+]}{d\,(pH)} = \frac{2.303\,[H^+] \times KC}{(K + [H^+])^2} = \beta \tag{7}$$

(See Van Slyke, 1922, p. 538).

Equations (4) and (7) have been plotted in Fig. 1 for a situation in which C = 20 mM and $K = 10^{-7}$, pK = 7. It will be noted that β is very far from constant except in the immediate vicinity of the pK.

Professor Piiper did mention in his address, I believe, that strictly speaking it was necessary for accuracy to integrate his equations. I should like to put forward the view that for our purposes it is better not to differentiate the fundamental equations in the first place.

D.R. Wilkie

Reference

SLYKE, R.D. van: J. Biol. Chem. 52, 525-570 (1922)

The well-known sigmoid buffer curve shown by Professor Wilkie applies to simple buffer systems involving a single H^+-dissociating substance or group. Blood (separated plasma, true plasma, or whole blood), however, is customarily treated in textbooks of physiology as exhibiting a straight buffer line, i.e., the buffer value β is considered to be practically independent of pH, in the physiological blood plasma pH range of 7.0-7.8. Very accurate measurements performed by Dill et al. (1) and by many others have shown this simplification to be quite acceptable for the analysis of acid-base relationships in blood.

The primary concern of my paper was buffering in the intracellular space. In experiments on rat diaphragm homogenates in my laboratory (2), β_{NB} was found to be practically constant in the pH range 6.3-7.1 (more precisely, on average a statistically insignificant increase by 8% was observed as pH was lowered from

7.1 to 6.3). Therefore, I believe that the use of a constant β, independent of pH, is sufficiently substantiated also for the analysis of intracellular acid-base balance.

This "broadening" of the buffer curve, or relative constancy of β, is usually attributed to the presence of multiple buffering substances and buffering groups and to modifications of the pK of buffering groups by the "molecular environment".

J. Piiper

References

1. DILL, D.B., EDWARDS, H.J., CONSOLAZIO, W.V.: Blood as a physicochemical system. Man at rest. J. Biol. Chem. 118, 635-648 (1937)
2. HEISLER, N., PIIPER, J.: The buffer value of rat diaphragm muscle tissue determined by P_{CO_2} equilibration of homogenates. Respir. Physiol. 12, 169-178 (1971)

Chapter 5

The Quantitive Significance of the "Himwich-Cori Cycle" for Removal of Lactate During Recovery After Maximal Exercise in Man

L. Hermansen and O. Vaage

Introduction

Berzelius was probably one of the first to discover that lactic acid was produced in the skeletal muscle during vigorous exercise. In his letter to Lehmann in 1841 (for reference see 1) he claimed to have found lactic acid in exhausted muscle. According to Bois-Reymond (1), Berzelius also stated that "ein Muskel desto mehr Milchsäure enthält, je mehr er vorher angestrengt worden ist" (the more a muscle has been exercised, the more lactic acid it contains). Later studies on skeletal muscle from different animal species and man (2, 3, 4, 5, 6) have supported and extended the findings of Berzelius. Thus, during maximal exercise of short duration there is a rapid production of lactate in the skeletal muscle. Some of the lactate produced in the muscle during exercise diffuses out into the blood and other fluid compartments. However, part of the lactate produced is "stored" within the muscle. Lactate concentrations of 25 $mmol \cdot kg^{-1}$ wet weight or more have been observed in human skeletal muscle after maximal exercise (4). Similar or somewhat lower lactate concentrations have been found in blood after maximal exercise (7, 8, 9, 10).

The present paper deals with the mechanism by which lactate is removed from skeletal muscle during recovery after maximal exercise.

Two separate lines of evidence have led to conflicting conclusions about the mechanism by which lactate is removed from skeletal muscle during recovery after exhaustive exercise. Meyerhof (5, 11) was the first to discover that some of the lactate produced during exercise was oxidised during the subsequent recovery period. However, only about 20%-25% of the total lactate disappearing from muscle could be accounted for by oxidation to carbon dioxide and water. Later studies by Hill (12) showed that the heat production of skeletal muscle during recovery was much smaller than (i.e., about one-quarter to one-fifth) would be expected if all the lactate produced was oxidised to carbon dioxide and water. Meyerhof (5) also showed that if the fatigued isolated muscle was allowed to recover in oxygen, glycogen was synthesised at the same time as lactate was disappearing. Thus he concluded that about one-quarter to one-fifth of the lactate produced during exercise was oxidised during the subsequent recovery period. The remaining lactate, i.e., three-quarters to four-fifths, was resynthesised to glycogen during the recovery period. These early results of Meyerhof (5, 11) were later confirmed by himself and his co-worker (13) and by others (2, 3, 14).

The opposing view is that lactate must first escape from the muscle into the general circulation, be carried to the liver, and there be converted to glucose. This glucose is then transferred to muscle, where it is ultimately converted to glycogen and stored. The evidence supporting this hypothesis came from the studies of Himwich et al. (15) and Cori and Cori (16). Himwich et al. (15), using decerebrated dogs, compared the concentrations of lactate and glucose in arterial blood with that in the venous blood draining the liver, skeletal muscles, heart, and other organs. They concluded that the main source of lactate under these conditions is the skeletal muscle, while the organ chiefly concerned with its removal is the liver. Moreover, since all the organs, with the exception of the liver, removed glucose from the blood while the liver added glucose, there seemed to be a carbohydrate cycle between muscle and liver, muscles sending lactic acid through the blood to the liver and the liver returning glucose to the muscle. Cori and Cori (16) extended these studies, and showed that sodium lactate, when given by mouth or by subentaneuns injection, led to a glycogen deposition in the liver. On the basis of their own results and those of Himwich et al. (15), Cori and Cori (16) emphasised the role of a complete cycle of the glucose molecule in the body.

Two other lines of evidence have been interpreted to support the Himwich-Cori cycle as the main mechanism by which lactate is removed from the muscle. First, skeletal muscle perfused or incubated with isotopically labelled lactate solutions has shown little or no increase in the level of isotopically labelled muscle glycogen (17, 18, 19). Second, Krebs and Woodford (20) and others (21, 22) have been unable to demonstrate the presence of all key enzymes necessary to convert lactate into glycogen in skeletal muscle. These enzymes were earlier found in the liver (20, 23).

However, recently the hypothesis put forward by Krebs and Woodford (20), that lactate cannot be resynthesised to glycogen in skeletal muscle has been severely criticised (2). Bendall and Taylor (2) confirmed the findings of Meyerhof (5, 11) showing that glycogen can be formed from lactate added to isolated frog muscle. These and other studies (2, 3, 8) have made it necessary to re-examine the whole problem of lactate removal from skeletal muscle.

In a recent series of experiments the question of whether or not lactate can be reconverted into glycogen in skeletal muscle tissue was investigated in our laboratory. Altogether five series of experiments were performed. The data presented in this paper have already been published elsewhere (4, 24).

Lactate Concentration and Glycogen Content in Human Skeletal Muscle During Recovery After Maximal Exercise

The objectives of the first set of experiments were:

1. To measure the rate of lactate disappearance in skeletal muscle during recovery after maximal exercise in man
2. To study the changes in the glycogen content of the same muscle during recovery after maximal exercise.

To ensure a partial depletion of the muscle glycogen stores and to produce a high concentration of lactate in the muscles, the protocol involving three bouts of intense exercise was adopted. The subjects (*n*=9) exercised three times at a work load (bicycle) leading to exhaustion within 60 s. A rest period of 4 min was allowed between each work bout. The work load was set individually in accordance with the performance during the preliminary testing 3-4 days before the actual experiment. After the third exercise bout, the subjects lay on a bed for 30 min. to recover. Muscle biopsy samples were taken while each subject was at rest before exercise, within 5-10 s after the end of the third exercise bout, and at 5, 10, 20, and 30 min during the recovery period. The muscle biopsy samples were taken from the lateral part of the quadriceps muscle by the needle biopsy technique described by Bergstrøm (25).

The values for muscle lactate concentrations at rest before maximal exercise and during the subsequent 30-min recovery period are given in Fig. 1 (lower panel). The mean ±SE for the muscle lactate con-

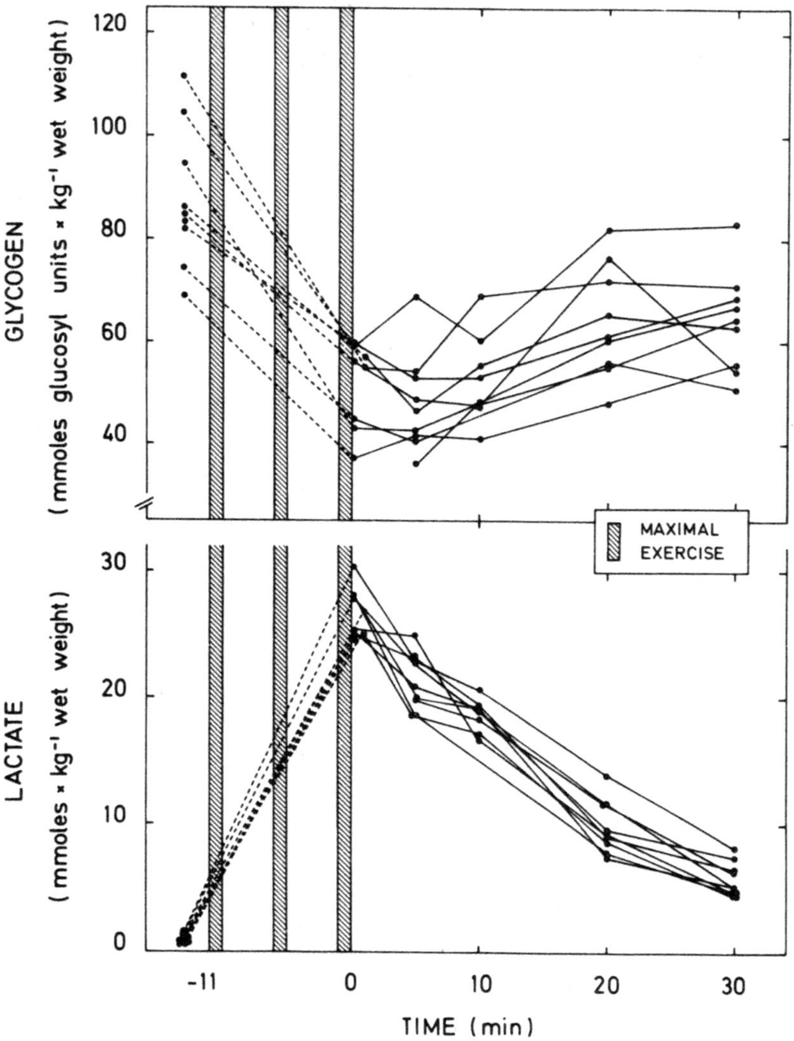

Fig. 1. Individual values for muscle glycogen content and muscle lactate concentration at rest before intermittent maximal exercise (▨) (i.e., three 1-min periods of exercise to exhaustion with 4-min rest periods in between) and during the first 30 min of recovery. L. Hermansen and O. Vaage, Am. J. Physiol. 233, E422-E429 (1977)

centration at rest was 1.1±0.2 $mmol \cdot kg^{-1}$ wet. Immediately after the third exercise bout, the muscle lactate concentration had increased to 26.4±1.0 $mmol \cdot kg^{-1}$ wet. During the 30-min recovery period the lactate concentration decreased almost linearly with time in all subjects. After 30 min of recovery the mean muscle lactate concentration had decreased to 6.6±0.9 $mmol \cdot kg^{-1}$ wet. Thus the average rate of lactate disappearance in human skeletal muscle during recovery after maximal exercise was found to be 0.66±0.02 $mmol \cdot min^{-1} \cdot kg^{-1}$ wet.

The changes in the muscle glycogen content during maximal intermittent exercise and in the subsequent recovery period are given in Fig. 1 (upper panel). At rest before the maximal intermittent exercise, the mean muscle glycogen concentration was 87.7±2.7 mmol glucosyl units · kg^{-1} wet. After the third maximal exercise bout, the mean muscle glycogen content had decreased to 49.6 mmol glucosyl units·kg^{-1} wet, i.e., the mean muscle glycogen concentration decreased by 38.1 mmol glycosyl units per kg muscle wet. In the subsequent recovery period there was a rapid increase in the muscle glycogen concentration (Fig. 1, upper panel). On average, the muscle glycogen content increased by 16.8 mmol glucosyl units per kg wet. Over the total recovery period (i.e., 30 min). Thus the average rate of glycogen synthesis was 0.56±0.09 mmol glucosyl units per kg muscle wet.

When these facts are taken together, it can be concluded that there is a rapid rate of lactate disappearance and glycogen synthesis during recovery after maximal exercise. Moreover, the disappearance of lactate and the glycogen synthesis coincide in time course and magnitude.

Changes in the Water Content of Muscle During Recovery After Maximal Exercise

In agreement with others (4, 26), we have found that the rate of lactate disappearance from muscle is rapid during recovery after maximal exercise. Our results also confirm an earlier observation showing an increase in muscle glycogen concentration during recovery after maximal exercise. However, our calculations of the rates of glycogen synthesis and lactate disappearance are based upon the assumption that the water content is constant during the recovery period. If the water content increased markedly during exercise, the estimated fall in muscle glycogen observed in our study would be too pronounced, due to a dilution effect. If one further assumes that all the extra water left the muscle during the recovery period, this would mean that our rate of glycogen synthesis has been overestimated. Therefore, we performed a new series of experiments, in which we measured the water content of the muscle before and after the maximal intermittent exercise.

The objective of this <u>second</u> set of experiments was to study the changes in water content of the human skeletal muscle in response to maximal intermittent exercise. The exercise and recovery protocols were the same as in the first set of experiments. These experiments involved six healthy subjects.

The wet weights were determined on a Cahn electrobalance at -23°C or below. The dry weights were determined on a Mettler belance to the nearest microgram after the tissue samples had been kept in small glass tubes at 105°C for 12 h or more. Protein was determined in residues after 3 M perchloric acid extraction and after the tissue had been dissolved in 0.7 N KOH at 70°C for 30 min by the biuret method.

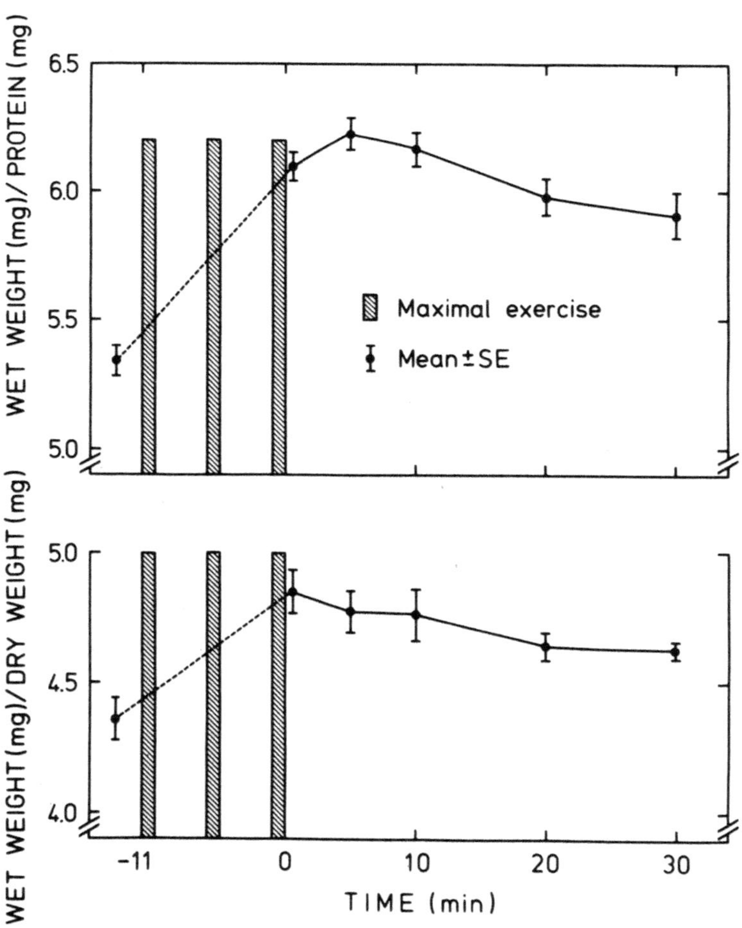

Fig. 2. Means ±SE (bars) for the wet weight:protein ratio (upper panel) and the wet weight:dry weight ratio at rest before intermittent maximal exercise (▨; see legend to Fig. 1) and during the following 30-min recovery period. L. Hermansen and O. Vaage, Am. J. Physiol. 233, E422-E429 (1977)

The mean value (±SE) for the wet weight: protein ratio (Fig. 2, upper panel) increased from 5.34±0.06 at rest before exercise to 6.10±0.05 immediately after the third exercise bout. This increase corresponded to a 14% increase in muscle water content. During the first 5 min of recovery the wet weight: protein ratio showed a further small increase to 6.23±0.06. From then on, a gradual decrease to 5.90±0.13 at the end of the recovery period was observed. Thus, the water content had increased to 114% immediately after the third exercise bout and decreased to 110% at the end of the recovery period, compared with the water content (100%) at rest before exercise.

The mean value (±SE) for the wet weight:dry weight ratio (Fig. 2, lower panel) increased from 4.36±0.08 at rest before exercise to 4.85±0.09 immediately after the third exercise bout. During recovery, the ratio decreased almost linearly with time, reaching 4.63±0.03 at the end of the recovery period. This corresponds to an increase in the water content to 111% immediately after exercise and a decrease to 106% at the end of recovery, compared with the resting water content of 100%. Thus, the results obtained with both these methods showed that the changes in the water content during the 30 min of recovery are small. Correction for these changes increased the calculated rate of lactate disappearance in the first set of experiments from 0.66 to 0.74 $mmol \cdot min^{-1} \cdot kg^{-1}$ wet and decreased the rate of glycogen synthesis from 0.56 to 0.51 mmol glucosyl units $\cdot$ $min^{-1} \cdot kg^{-1}$ wet.

In Fig. 3 some of the above results are replotted to show the changes in the water content of muscle in response to maximal exercise. In the upper part of the figure, changes in the dry weight and water are compared, with the wet weight of the samples kept constant. The lower part of the figure shows changes in the water content of the muscle while the dry weights are constant.

Thus, it can be concluded that the water content of the muscle increases during maximal intermittent exercise and that part of this extra water leaves the muscle during the first 30 min of recovery. However, the changes in the water content of muscle during recovery after maximal exercise (i.e., the first 30 min) are small, and affect the calculations of the rates of glycogen synthesis and lactate disappearance only to a minor extent.

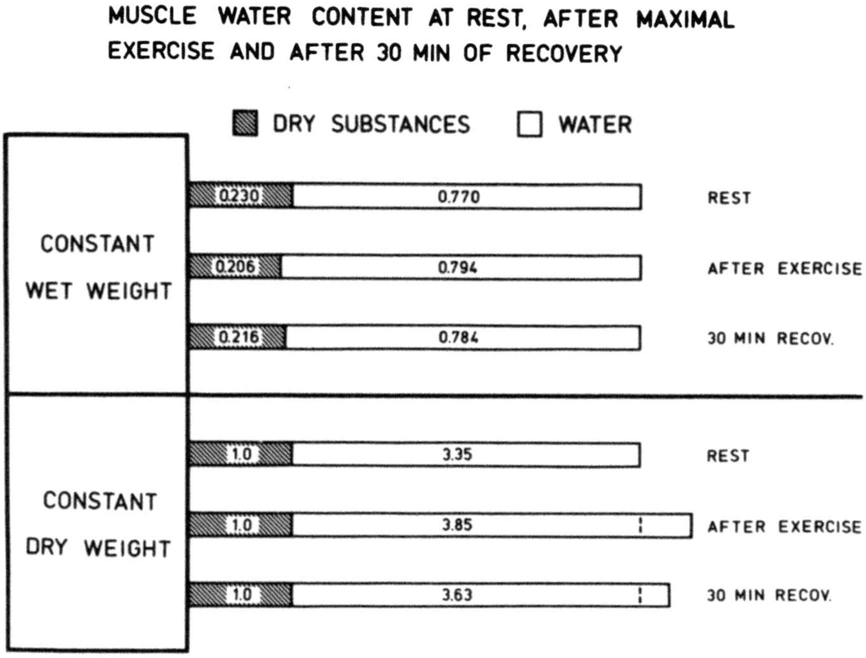

Fig. 3. A schematic illustration of the changes in the water content of muscle. In the upper part of the figure the wet weights are kept constant, while in the lower part the dry weights are constant. ▨, dry substances; □, water

Exchange of Lactate, Glucose, and Alanine Between Blood and Muscle in Response to Maximal Exercise

In the first set of experiments it was shown that there is a rapid disappearance of lactate and a rapid synthesis of glycogen in skeletal muscle during recovery after maximal exercise (see Fig. 1). These observations raised two important questions. First, the fate of the removed lactate; and second, the source of the glycogen. It was to answer these questions that the third set of experiments was performed. The objective of these experiments was to determine the exchange of lactate, glucose, and alanine between blood and muscle at rest before exercise and during recovery after maximal exercise. In addition, the changes in the plasma immunoreactive insulin (IRI) were determined.

About 30 min before these experiments began, a 20-gauge polyethylene catheter was inserted percutaneously about 10 cm into a femoral artery (A), and another was inserted into a femoral vein (FV). The subjects (n=4) this time performed maximal exercise on a motor-driven treadmill with an inclination of 3° (5.25%). Otherwise, the exercise and recovery protocols were the same as for the first set of experiments. Blood samples were drawn simultaneously from the femoral artery and the femoral vein at rest before the first maximal work bout, within 5-15 s after the third exercise bout, and at 5, 10, 2o and 30 min during the recovery period. The blood samples taken immediately after the third exercise bout were drawn while the subjects were standing: the remainder were taken with subjects lying supine on the bed.

Intermittent maximal exercise increased the arterial blood lactate concentration (Fig. 4) from 1.3±0.3 mmol·litre^{-1} at rest before exercise to 20.9±1.2 mmol·litre^{-1} (mean value ±SE) after the third exercise bout. The arterial blood lactate concentration stayed at this level for approximately 5 min. From then on, an almost linear decrease was observed at an average rate of 0.43±0.05 mmol·min^{-1}·litre^{-1}. After 30 min of recovery, the arterial blood lactate concentration was still 10.6±1.1 mmol·litre^{-1}. The femoral venous blood lactate concentrations were only slightly higher than those in the corresponding arterial blood samples during the entire recovery period (Fig. 5, lower panel). The mean A-FV lactate difference (i.e., for the whole recovery period) was 0.58±0.13 mmol·litre^{-1}, and no statistically significant changes were observed in A-FV lactate difference during recovery (F = 0.23).

The mean arterial blood glucose concentration (Fig. 4) increased from 7.3±1.3 mmol·litre^{-1} at rest to 8.5±0.6 mmol·litre^{-1} immediately after the intermittent exercise. An additional increase to 8.9±0.6 mmol·litre^{-1} was observed during the first 5 min of recovery. From then on, a linear decrease toward pre-exercise values was found during the last 25 min of recovery. The A-FV glucose differences (Fig. 5) were small and showed no statistically significant changes during the recovery period. The mean value for the A-FV glucose differences over the whole recovery period was 0.18±0.03 mmol·litre^{-1}. The mean ±SE for the plasma IRI concentration (Fig. 4) increased from 28.4±15.6 μIU·ml^{-1} at rest before exercise to 35.8±5.9 μIU·ml^{-1} immediately after the third exercise bout. During the first 5 min of recovery, an additional pronounced increase in the plasma IRI concentration was observed, followed by a decrease to 19.5±6.1 μIU·ml^{-1} during the last 20 min of the recovery period. No A-FV plasma IRI differences were observed during recovery.

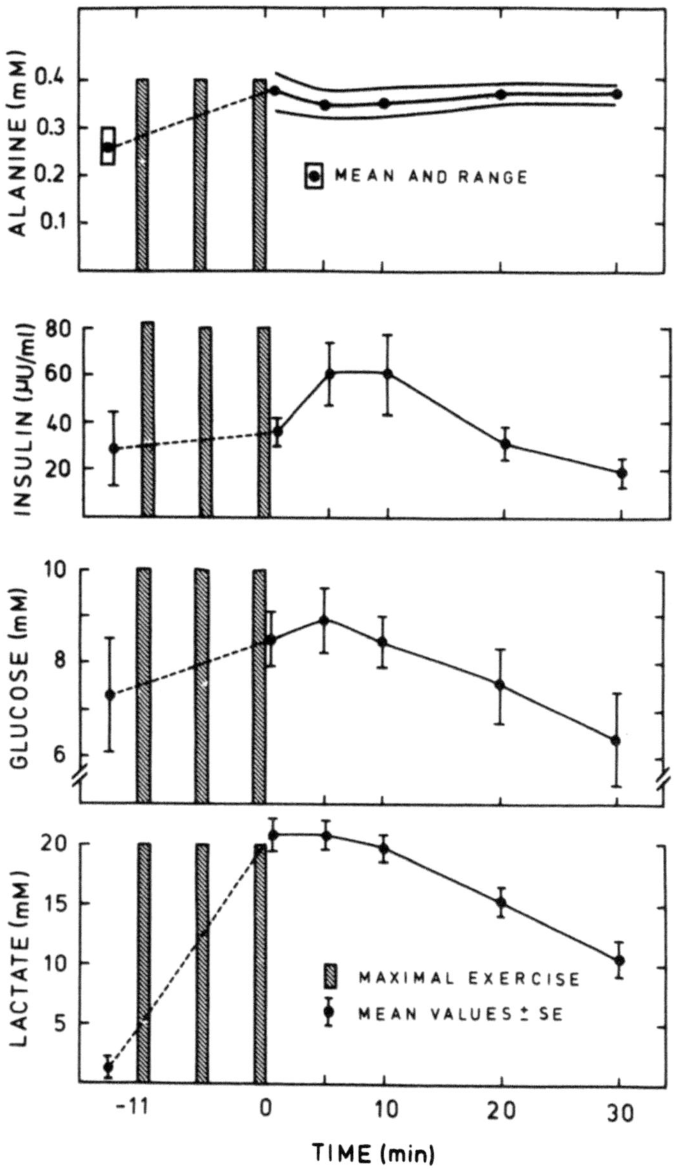

Fig. 4. Means ±SE (ɸ) or mean and range (▣) for the concentrations of alanine, insulin, glucose, and lactate in arterial blood (plasma for insulin) at rest before intermittent maximal exercise (▨; see legend to Fig. 1) and during the subsequent recovery period (30 min). L. Hermansen and O. Vaage, Am. J. Physiol. 233, E422-E429 (1977)

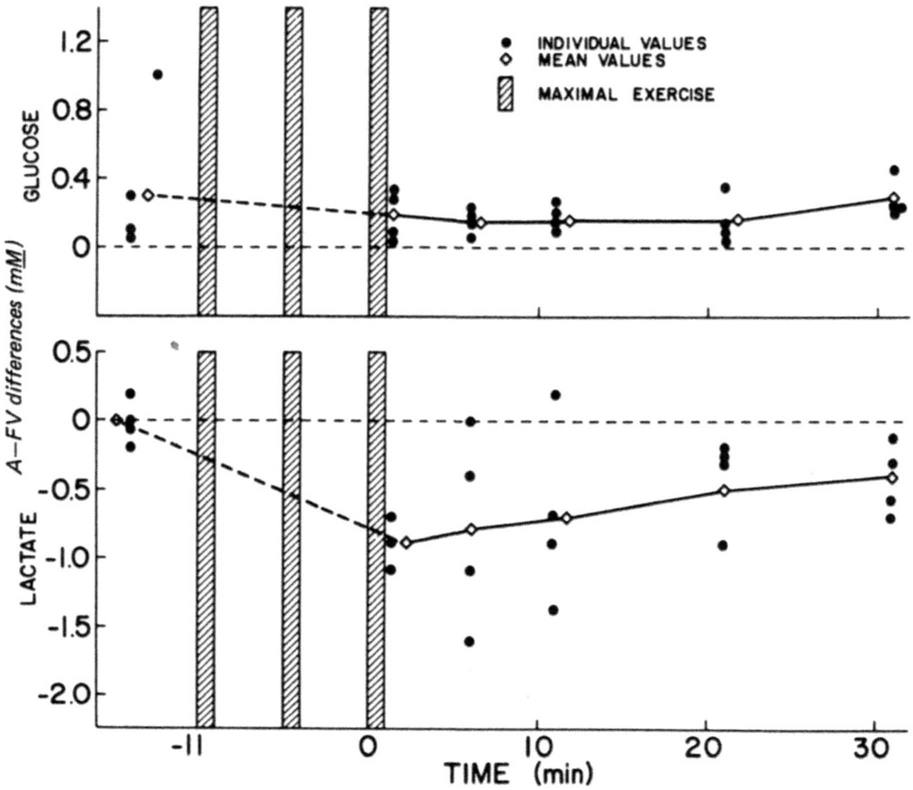

Fig. 5. Individual (●) and mean (◇) values for A-FV glucose and lactate differences at rest before intermittent maximal exercise (▨; see legend to Fig. 1) and during the first 30 min of recovery. L. Hermansen and O. Vaage, Am. J. Physiol. 233, E422-E429 (1977)

In contrast to the relatively large changes observed in the arterial blood concentrations of lactate and glucose and in the plasma IRI concentrations during and after intermittent maximal exercise, blood alanine concentrations (Fig. 4) increased only from 0.27 $mmol \cdot litre^{-1}$ at rest to 0.37 $mmol \cdot litre^{-1}$ immediately after the third exercise bout. Blood alanine concentrations remained unchanged throughout the recovery period, and no A-FV alanine differences were found.

Thus, according to the results obtained in the third set of experiments only small A-FV differences for lactate, glucose, and alanine were observed. However, values for muscle blood flow are needed to calculate the amount of lactate released from muscle to blood, or the amount of glucose taken up by muscle.

Blood Flow in Skeletal Muscle During Recovery After Maximal Exercise

The object of the fourth set of experiments was to study the recovery of heart rate and total calf blood flow (CBF) (27) (venous plethysmography) after intermittent maximal exercise to approximate the changes in muscle blood flow. The exercise protocol was exactly the same as in the first set of experiments. Before the exercise began, cuffs were placed just above the ankle and just below the knee. Immediately after the third exercise bout, the subjects (n=5) moved quickly to a bed. One leg was supported above venostatic level by a sling at the ankle. The plethysmography gauge was then quickly (usually within 1 min) placed around the calf. CBF decreased from 21 to 9 $ml \cdot min^{-1} \cdot 100\ ml^{-1}$ muscle plus skin tissue from the first to the fifth minute of recovery (Fig. 6, lower panel). Thereafter, CBF remained at

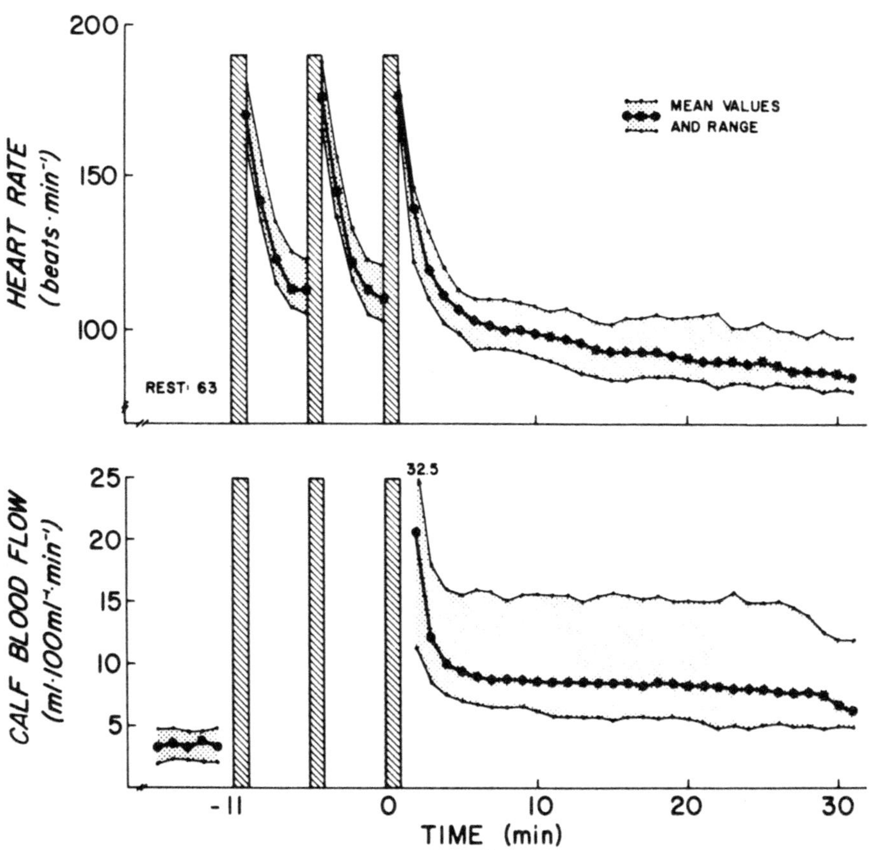

Fig. 6. Mean values (-•-•-) and range (-·-·-) for heart rate and total CBF at rest before intermittent maximal exercise (▨; see legend to Fig. 1) and during the subsequent 30-min recovery period. L. Hermansen and O. Vaage, Am. J. Physiol. 233, E422-E429 (1977)

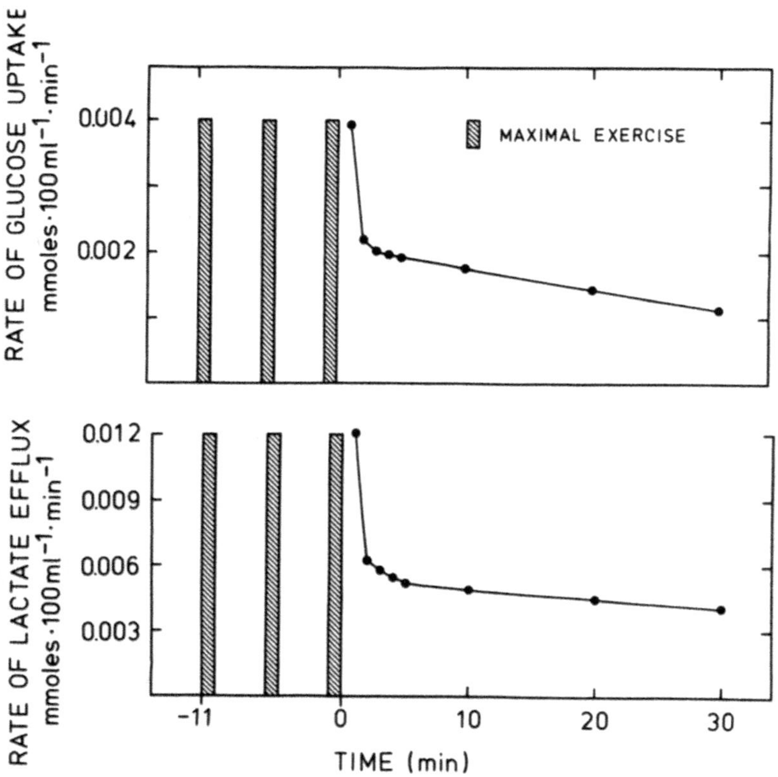

Fig. 7. Mean values for the rates of lactate efflux from the muscle (lower panel) and its uptake of glucose (upper panel) during recovery after maximal exercise (▨; see legend to Fig. 1). L. Hermansen and O. Vaage, Am. J. Physiol. 233, E422-E429 (1977)

approximately double the pre-exercise control level. The mean heart rate (Fig. 6, upper panel) fell from 176 to 103 beats·min^{-1} during the first 5 min of recovery and declined slowly thereafter, indicating that the cardiac output dropped rapidly during the first few minutes of recovery, and from then on stayed at a relatively constant level somewhat higher than the level at rest.

On the basis of the results obtained in the third and fourth sets of experiments, the substrate fluxes were calculated. The rate of lactate efflux from muscle was calculated to be 0.012 mmol·min^{-1}· 100 ml^{-1} 1 min after the third bout of exercise (Fig. 7, lower panel). It decreased rapidly during the next 4 min of recovery, due to the rapid fall in muscle blood flow. During the last 25 min of recovery lactate efflux decreased slowly to 0.004 mmol·min^{-1}·100 ml^{-1}.

Taken as a whole, lactate efflux from the muscle accounted for less than 10% of the lactate disappearance from muscle during the recovery period.

The rate of glucose uptake, based on the measurements of A-FV glucose differences and CBF, was 0.004 $mmol \cdot min^{-1} \cdot 100\ ml^{-1}$ 1 min after the end of the third exercise bout (Fig. 7, upper panel) and decreased rapidly during the next 4 min. During the last 25 min of recovery, glucose uptake decreased from 0.0019 to 0.0012 $mmol \cdot min^{-1} \cdot 100\ ml^{-1}$. Thus, glucose uptake accounted for about 3% of the muscle glycogen synthesis during the 30-min recovery period.

With the above results in mind, we will go back to the two questions mentioned earlier; first, what is the fate of the large amount of lactate that is removed so quickly from the muscle? Part of the muscle lactate is cleared to the blood. However, as shown in Fig. 5, the A-FV lactate difference was very small. Thus, to account for the lactate disappearance in muscle through a removal of lactate via the circulation, leg blood flow would have to average 20-25 $litres \cdot min^{-1}$. A value of this order corresponds to the total cardiac output during maximal exercise. From our measurements of A-FV lactate differences and CBF, we have calculated that less than 10% of the muscle lactate is cleared to the blood during the recovery period. Consequently, more than 90% (i.e., 0.67 $mmol \cdot min^{-1} \cdot kg^{-1}$ wet) must be metabolised in muscle itself.

Because lactate is a metabolic "dead end", its most probable fate in skeletal muscle is conversion back to pyruvate. The necessary enzyme, lactate dehydrogenase, and its appropriate isoenzymes are present in human skeletal muscle (28). After conversion to pyruvate, there are at least two possible fates: (1) oxidation to carbon dioxide and water, and (2) conversion to alanine.

Meyerhof (5, 11) and Hill (12) found that less than 25% of the lactate in isolated amphibian skeletal muscle was oxidised to carbon dioxide and water. According to our measurements of CBF and measurements of Keul and co-workers (29) of A-FV oxygen differences under similar conditions, about 15% of the muscle lactate disappearance might be attributable to oxidation. Another way of evaluating muscle lactate oxidation is to calculate the A-FV oxygen difference that would be needed to provide sufficient oxygen for oxidation. Three moles of oxygen are needed to oxidise 1 mol lactate. Thus the A-FV oxygen difference at the CBF we measured would have to exceed 45 ml oxygen per 100 ml blood to account for oxidation of the muscle lactate. This is about three times higher than the A-FV oxygen difference observed during maximal exercise, and ten times higher than the A-FV oxygen difference observed by Keul et al. (29) under similar conditions. This physiological impossibility leads to the conclusion that only a fraction of the lactate disappearing in muscle is oxidised.

In our calculations of lactate and glucose fluxes, we assumed that the Fick principle was applicable. This may not be correct for the first few minutes of the recovery period. During the last 25 min of the recovery period, CBF was sufficiently constant to justify use of the Fick principle (Fig. 6). The arterial concentrations of glucose and lactate changed during the recovery period. The A-FV differences given in Fig. 5 were obtained from measurements in arterial and venous blood samples taken simultaneously, whereas the true A-FV difference is between the venous concentration and some earlier arterial concentration. This would make the true A-FV differences (Fig. 5) for lactate somewhat lower and those for glucose somewhat higher. Thus, our calculated value for lactate efflux is probably an overestimate.

Another possibility for metabolism of lactate in skeletal muscle is utilisation of the "alanine cycle", as proposed by Felig and co-workers (30, 31) and Malette et al. (32). Lactate (via pyruvate) is converted to alanine by the glutamate pyruvate transaminase reaction. However, we found that the concentration of alanine in the blood increased by only 0.1 $mmol \cdot litre^{-1}$ during the intermittent maximal exercise, and no A-FV alanine differences were observed during the recovery period. Therefore, the alanine cycle appears to be of minor quantitative significance in accounting for the disappearance of lactate from skeletal muscle under the conditions of these experiments.

Of the 0.74 mmol lactate $\cdot$ kg^{-1} wet that disappears in the muscle each minute, less than 10% can be accounted for by an efflux to the blood, and about 15% is accounted for by oxidation to carbon dioxyde and water. This means that 75% (0.56 $mmol \cdot min^{-1} \cdot kg^{-1}$ wet) must be metabolised by some other route (or routes) within the muscle.

The second interesting question is that of the identity of possible precursors for the rapid synthesis of glycogen in the muscles. Our results confirm earlier observations showing an increase in the muscle glycogen content during recovery after maximal exercise. Meyerhof (5, 11) found an increase in the glycogen content of isolated amphibian skeletal muscle if the muscle was allowed to recover in oxygen. The same observation has been made by Bendall and Taylor (2) in the skeletal muscle of frogs and rabbits and by Karlsson and Saltin (26) in man. However, the significance of these observations has been questioned. First, the rates of glycogen synthesis observed in isolated muscle preparations are very low (i.e., about 0.08 mmol glucosyl units$\cdot$min$^{-1}\cdot$kg^{-1} wet wt.). Second, in the studies of intact organisms in which an increase in muscle glycogen was observed during recovery after maximal exercise (26), no measurements of glucose uptake were made. Thus, the glycogen synthesis might have been accounted for by an uptake of glucose from the blood.

We observed a very rapid glycogen synthesis, i.e., 0.51 mmol glucosyl units$\cdot$min$^{-1}\cdot$kg^{-1} wet. This rate is four times as fast as the rate of glycogen synthesis observed in normal subjects on a carbohydrate-rich diet during recovery after prolonged exercise that had essentially depleted the muscle glycogen stores (33). Moreover, the rate of glycogen synthesis is related to the glycogen content of the muscle (33). If we compare the rate of glycogen synthesis after prolonged exercise that produces only partial glycogen depletion, i.e., at about the same level as in the present study, our rate of glycogen synthesis is approximately eight times faster.

Where does the glycogen come from? Our results indicate that some of the glycogen is synthesised from blood glucose. According to our measurements of CBF and A-FV glucose differences, only about 0.6 of the 15 mmol glucosyl units$\cdot$kg^{-1} wet synthesised during the 30 min of recovery is attributable to an uptake of glucose from the blood. In contrast to the situation discussed above for lactate, any circulatory lag would tend to make the A-FV glucose difference larger. We have calculated that the true A-FV glucose differences are probably about 50% higher than those given in Fig. 5. Even with these corrections, glucose uptake may account for no more than 5% of the glycogen synthesised.

The small uptake of glucose cannot be explained by lack of available substrate or insulin. Blood glucose and plasma IRI concentrations were higher than resting values throughout the recovery period. However, intracellular factors may also influence the glucose uptake by the muscle cell. The increase in blood glucose that we found is consistent with earlier observations from this laboratory (34), and is probably explained by liver glycogenolysis as a result of increased release of catecholamines during the maximal exercise. The rise in plasma IRI is probably a secondary response to the increase in blood glucose concentration. Although the pronounced increase in the plasma IRI seems to have little effect on the glucose uptake by muscle, it may be an important factor for the rapid synthesis of glycogen observed in the present investigation. Gourley and Suh (35) showed that glycogen synthesis in frog sartorius muscle from either glucose or lactate was stimulated by insulin. Moreover, insulin not only facilitates glucose transport, but also stimulates the activity of glycogen synthetase. An increase in glycogen synthetase activity is probably necessary to account for the rapid synthesis of glycogen observed in the present investigation.

In conclusion, we have confirmed earlier observations (26) showing a rapid disappearance of lactate from human skeletal muscle during its recovery from maximal exercise. Only a small fraction (about 10%) of this lactate can be accounted for by an efflux from the muscle into the circulation. The major fraction (90%) must be metabolised within muscle tissue. Less than 15% of the lactate disappearing from muscle during recovery appears to be oxidised to carbon dioxide and water. The remainder, about 75%, must be metabolised by another route (or routes). A strong possibility is conversion to glycogen, on the basis of the following evidence: (1) the rate of glycogen synthesis during recovery was rapid; (2) only a small fraction of the glycogen synthesised could be accounted for by an uptake of glucose from the blood, so that more than 90% of the glycogen must have been synthesised from other precursors; (3) disappearance of lactate and synthesis of glycogen during recovery after maximal exercise coincide both in time course and in magnitude; (4) neither the Himwich-Cori cycle nor the alanine cycle can account for either the lactate disappearance or the glycogen synthesis.

Muscle Metabolites During Recovery After Maximal Exercise

The results given above have shown that only a small fraction of the lactate produced in skeletal muscle during maximal exercise disappeared from muscle into circulation during recovery. This observation and our findings of a rapid lactate disappearance in muscle, a rapid muscle glycogen resynthesis, and almost no uptake of glucose

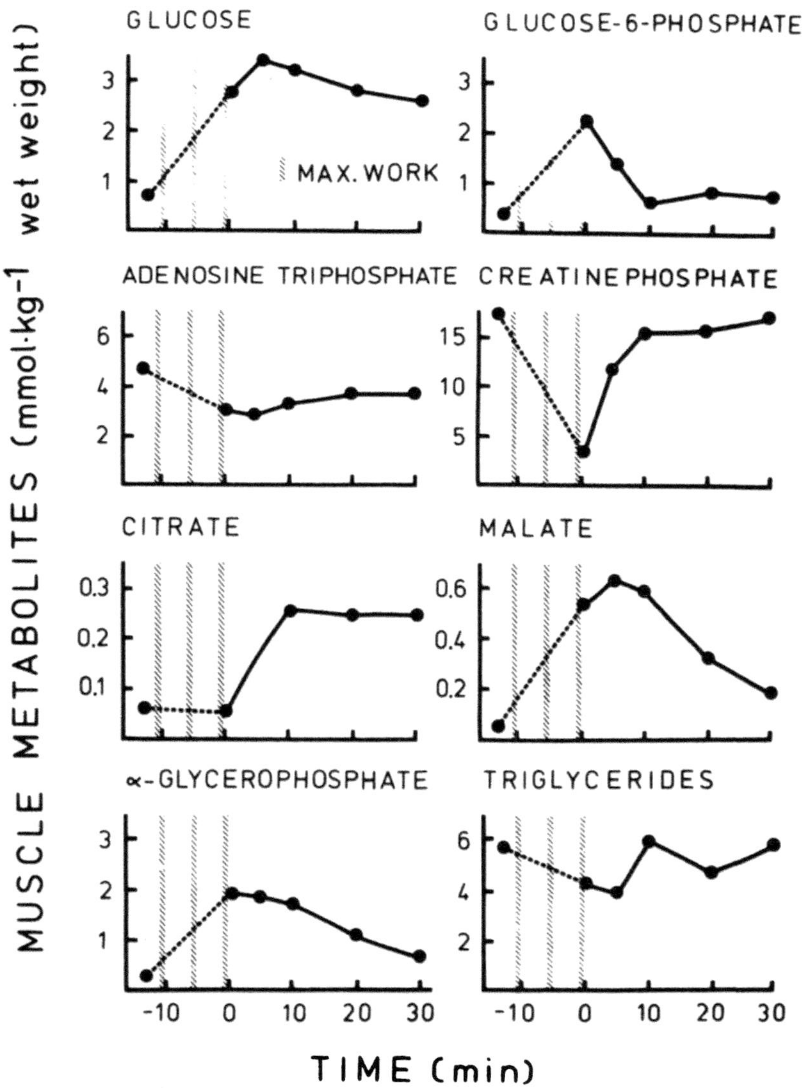

Fig. 8. Mean values for concentration of different muscle metabolites at rest before maximal intermittent exercise (▨; see legend to Fig. 1) and during the subsequent 30-min recovery period. O. Vaage et al., Acta Physiol. Scand. 102, A11-A12 (1978)

by the muscles during recovery, were offered in support of Meyerhof's (5) original hypothesis of a conversion of lactate into glycogen in skeletal muscle. To test this hypothesis further, a fifth set of experiments was performed, in which changes in the concentrations of adenosine triphosphate (ATP), creatine phosphate (CP), triglycerides, and different glycolytic intermediates were measured in the quadriceps muscle of 15 healthy male subjects at rest before intermittent maximal exercise and at different times during the recovery period (30 min), i.e., the same protocol as in the first set of experiments. The results (Fig. 8) indicated that changes in the glycolytic intermediates could account for only a small fraction of the glycogen synthesised during the recovery period. The glucose and glucose-6-phosphate concentrations during recovery are consistent with an inhibition of hexokinase and with the low uptake of glucose by muscle reported above. Moreover, the concentrations of ATP, CP, and citrate observed during recovery suggest that phosphofructokinase is also inhibited. Altogether, the results of these experiments are consistent with the hypothesis of a conversion of lactate into glycogen in human skeletal muscle.

Summary

The present paper deals with the problem of lactate removal during recovery after maximal exercise in man. The data presented show that only a small fraction (less than 10%) of the lactate disappearing from skeletal muscle during recovery after maximal exercise can be accounted for by a diffusion of lactate from the muscle into the circulation. Moreover, glucose uptake by the muscle is very low, i.e., less than 5% of the glycogen synthesised in the muscles. Thus it can be concluded that the significance of the Himwich-Cori cycle for the removal of lactate from muscle during recovery after maximal exercise is very small. Our data support the original hypothesis of Meyerhof (23), who suggested that a large fraction (about 75%) of the lactate produced during maximal exercise is resynthesised to glycogen in the muscles.

References

1. DU BOIS-REYMOND, E.: Über angeblich saure Reaktion des Muskelfleisches. In: Gesammelte Abhandlungen zur allgemeinen Muskel- und Nervenphysik. 1877
2. BENDALL, J.R., TAYLOR, A.A.: The Meyerhof quotient and the synthesis of glycogen from lactate in frog and rabbit muscle. A reinvestigation. Biochem. J. 118, 887-893 (1970)
3. FOSTER, D.L., MOYLE, D.M.: A contribution to the study of the interconversion of carbohydrate and lactic acid in muscle. Biochem. J. 15, 672-680 (1921)
4. KARLSSON, J.: Lactate and phosphagen concentrations in working muscle of man. Acta Physiol. Scand. [Suppl.] 358, 72 (1971)
5. MEYERHOF, O.: Über die Energieumwandlungen im Muskel. II. Das Schicksal der Milchsäure in der Erholungsperiode des Muskels. Pfluegers Arch. 182, 284-317 (1920)
6. STAINSBY, W.H., WELCH, H.G.: Lactate metabolism of contracting dog skeletal muscle in situ. Am. J. Physiol. 211, 177-183 (1966)
7. HERMANSEN, L.: Lactate production during exercise. In: Muscle Metabolism during Exercise. Pernow, B., Saltin, B. (eds.), pp. 401-407. New York: Plenum Press 1971
8. HERMANSEN, L., STENSVOLD, I.: Production and removal of lactate during exercise in man. Acta Physiol. Scand. 86, 191-201 (1972)
9. OSNES, J.-B., HERMANSEN, L.: Acid-base balance after maximal exercise of short duration. J. Appl. Physiol. 32, 59-63 (1972)
10. STROM, G.: The influence of anoxia on lactate utilization in man after prolonged work. Acta Physiol. Scand. 17, 440-451 (1949)
11. MEYERHOF, O.: Die Energieumwandlungen im Muskel. VI. - Über den Ursprung der Kontraktionswärme. Pfluegers Arch. 195, 22-74 (1922)
12. HILL, A.V.: Muscular activity and carbohydrate metabolism. Science 60, 505-514(1924)
13. MEYERHOF, O., LOHMANN, K., MEIER, R.: Über die Synthese des Kohlehydrats im Muskel. Biochem. Z. 157, 459-491 (1925)
14. HERMANSEN, L., VAAGE, O.: Lactate disappearance and glycogen in human muscle after maximal exercise. Am. J. Physiol. 233 (5), E422-E429 (1977)
15. HIMWICH, H.E., KOSKOFF, Y.D., NAHUM, L.H.: Studies in carbohydrate metabolism. l.A. glucoselactic acid cycle involving muscle and liver. J. Biol. Chem. 85, 571-584 (1929/30).
16. CORI, C.F., CORI, G.R.: Glycogen formation in the liver from d- and l-lactic acid. J. Biol. Chem. 81, 389-403 (1929)
17. HIATT, H.H., GOLDSTEIN, M., LAREAU, J., HORECKER, B.: The pathway of hexose synthesis from pyruvate in muscle. J. Biol. Chem. 231, 303-307 (1958)
18. OMACHI, A., LIFSON, N.: Metabolism of isotopic lactate by the isolated dog gastrocnemius. Am. J. Physiol. 185, 35-40 (1956)
19. WARNOCK, L.G., KOEPPE, R.E., INCIARDI, N.F., WILSON, W.E.: L (+) and D (-) lactate as precursors of muscle glycogen. Ann. N.Y. Acad. Sci. 119, 1048-1060 (1965)
20. KREBS, H.A., WOODFORD, M.: Fructose 1,6-diphosphatase in striated muscle. Biochem. J. 94, 436-445 (1965)
21. CRABTREE, B., HIGGINS, S.J., NEWSHOLME, E.A.: The activities of pyruvate carboxylase, phosphoenolpyruvate carboxylase and fructose diphosphatase in muscle from vertebrates and invertebrates. Biochem. J. 130, 391-396 (1972)
22. OPIE, L.H., NEWSHOLME, E.A.: The activities of fructose 1,6-diphosphatase, phosphofructokinase, phosphoenolpyruvate carboxykinase in white and red muscle. Biochem. J. 103, 391-399 (1967)
23. KREBS, H.A.: Gluconeogenesis. Proc. R. Soc. Lond. 159, 545-565 (1964)
24. VAAGE, O., NEWSHOLME, E., GRØNNERØD, O., HERMANSEN, L.: Muscle metabolites during recovery after maximal exercise in man. Acta Physiol. Scand. 102, A11-A12 (1978)
25. BERGSTRØM, J.: Muscle electrolytes in man. Scand. J. Clin. Lab. Invest. [Suppl.] 68, 110 (1962)
26. KARLSSON, J., SALTIN, B.: Oxygen deficit and muscle metabolites in intermittent exercise. Acta Physiol. Scand. 82, 115-122 (1971)
27. BRENGELMANN, G.L.C., WYSS, ROWELL, L.B.: Control of forearm skin blood flow during periods of steadily increasing skin temperature. J. Appl. Physiol. 35, 77-84 (1973)

28. SJODIN, B.: Lactate dehydrogenase in human skeletal muscle. Acta Physiol. Scand. [Suppl.] 436 (1976)
29. KEUL, J., DOLL, E., KEPPLER, D.: Muskelstoffwechsel. p. 247. München: Barth 1969
30. FELIG, P., POZEFSKY, T., MARLISS, E., CAHILL, G.F., ALANINE, J.R.: Key role in gluconeogenesis. Science 167, 1003-1004 (1970)
31. WAHREN, J., FELIG, P., HENDLER, R., AHLBORG, G.: Glucose and amino acid metabolism during recovery after exercise. J. Appl. Physiol. 34, 838-845 (1973)
32. MALLETTE, L.E., EXTON, J.H., PARK, C.R.: Control of glyconeogenesis from amino acids in perfused rat liver. J. Biol. Chem. 244, 5713-5723 (1969)
33. PIEHL, K.: Time course for refilling of glycogen stores in human muscle fibres following exercise-induced glycogen depletion. Acta Physiol. Scand. 90, 297-302 (1974)
34. HERMANSEN, L., PRUETT, E.D.R., OSNES, J.-B., GIERE, F.A.: Blood glucose and plasma in response to maximal exercise and glucose infusion. J. Appl. Physiol. 29, 13-16 (1970)
35. GOURLEY, D.R.H., SUH, T.K.: Effects of insulin on oxidation and glyconeogenesis from glucose and glucose plus lactate in frog skeletal muscle. Comp. Biochem. Physiol. 29, 137-148 (1969)

Chapter 6

"Oxyen Dept" Does Not Exist

P. Harris

Early experiments on frog muscles showed that contraction could occur in the absence of oxygen. Fletcher and Hopkins (1) showed that excised muscle developed lactate when kept under anaerobic conditions, but not under aerobic conditions. Fatigue due to contractions of excised muscle was accompanied by an increase in lactate; but, if the fatigued muscle was placed in oxygen, the lactate disappeared. Fletcher and Hopkins suggested that the energy for contraction was derived from the liberation of lactate which, in the presence of oxygen could be oxidised or restored to its pre-existing chemical state.

The experiments of Hill (2, 3) showed that the "initial heat" of contraction was independent of the presence of oxygen, while the "recovery heat" required oxygen. However, Hill (4) calculated that the recovery heat accounted for only one-fifth of the heat of combustion of lactate and proposed that only some of the lactate was oxidised during recovery, the rest being resynthesised to its precursor.

Meyerhof (5, 6, 7) identified the precursor as glycogen. After stimulation of excised muscle, glycogen disappeared. If the muscle was kept in oxygen, glycogen was resynthesised.

It was against this background that the theory of "oxygen debt" developed. According to the original theory (8), the increased uptake of oxygen following exercise was due to the oxidation of lactate which, thereby, provided the energy for the resynthesis of glycogen.

All subsequent history has been a retreat from this position. In 1933, Margaria et al. (9) found that after light exercise there need be no increase in the concentration of lactate in the blood, although a "recovery volume" or "oxygen debt" of up to 4 litres could be measured. The observation came shortly after the discovery of creatine phosphate and Margaria and his colleagues suggested that, in addition to Hill's "lactacid debt," there must be an "alactacid" debt.

Other authors (10, 11) have shown that, during mild prolonged exercise, the blood lactate rises to an early maximum and falls again independently of the oxygen uptake. During brief periods of mild exercise the blood lactate rises to a maximum after the end of exercise and persists elevated after the oxygen uptake has returned to its resting value.

Huckabee (12, 13) showed that the blood lactate could increase during the infusion of glucose or bicarbonate without any change in oxygen uptake. He suggested that the ratio of lactate to pyruvate had to be taken into account and devised the calculation of "excess lactate" which he could relate to the oxygen debt. The calculation of excess lactate was a mathematical absurdity, since it started from the premise that the concentration of lactate was a function of the product of two quantities and ended with an equation which expressed it as a function of their addition (14). Moreover, in the hands of other workers, it could be shown to reach a negative value during prolonged exercise (11).

The original contention that part of the lactate produced by exercise goes to reform glycogen in the muscles is no less doubtful. Hultman (15) has shown that, in man, muscle glycogen can take several days to recover if the subject is kept on a carbohydrate-free diet. Very little labelled lactate, infused into the blood-stream, is incorporated in muscle glycogen (16). The enzymes necessary for the reversal of glycolysis are present to only a small degree in red muscle. Finally, careful perusal of Meyerhof's original papers shows that he studied amphibian muscle and measured recovery glycogen only after the muscle had been left for 23 h in oxygen.

In view of this evidence, it seems clear that the concentration of lactate in the blood does not play any causative role (1) in the increased uptake of oxygen after exercise (2) in the resynthesis of muscle glycogen after exercise. It is astounding that such notions should still persist after the biochemical advances of the last half century.

An important element of the original oxygen debt theory was the analysis of the shape of the relation between oxygen uptake and time. Although the achievement of a steady rate of external work is, for practical purposes, instantaneous, the oxygen uptake at the mouth takes 3 or 4 min to reach a new steady state, even during mild or moderate exercise. Conversely the oxygen uptake at the mouth remains high for a period of time after exercise.

An important distinction had to be made between the oxygen consumption and the oxygen requirement during exercise. "In the case of moderate exercise", said Hill et al. (18), "the oxygen intake gradually rises, as the exercise is continued, until in two or three minutes it equals the oxygen requirement: the recovery oxygen in this case is simply equal to the initial lag in the oxygen intake. In the case, however, of more severe exercise, the oxygen intake can never catch up with the oxygen requirement, and a large oxygen debt is incurred which is paid off during recovery."

For the present discussion, I propose to avoid using the term "oxygen debt", preferring to use terms which are descriptive rather than interpretative. They are indicated in the diagram in Fig. 1.

This form of diagram has now become so commonplace that there can scarcely be a textbook of physiology which does not show it. I believe the prototype is due to Krogh and Lindhard, who in 1920 (19) studied one subject on one occasion and found that the "lag volume" was

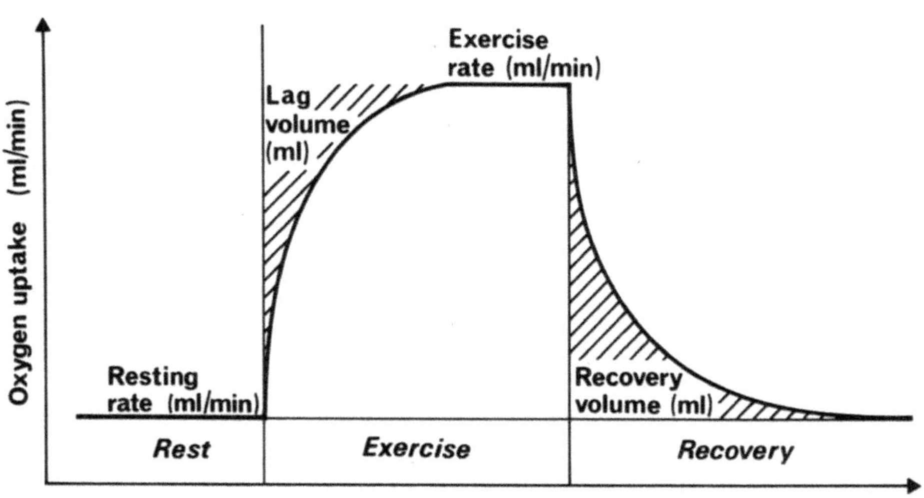

Fig. 1. Definition of terms used to describe the change in oxygen uptake during exercise and recovery

approximately equal to the recovery volume. Two years later, Lupton (20) said that Krogh and Lindhard had found the two areas to be the same. Hill and his colleagues, as we have seen, accepted it as fact for moderate exercise, and over the years this view has become established. The truth, however, is otherwise. A number of physiologists have investigated the relative magnitude with care (21, 22, 23) and the recovery volume almost always exceeds the lag volume, even at low levels of exercise.

It might be argued that the requirement for oxygen normally exceeded the rate of uptake even during the steady state of moderate exercise, so that the real deficit represented a larger area than that of the lag volume. If this were so, the debt could accrue throughout exercise and hence the recovery volume would increase with an increasing length of steady-state exercise. We now know, however, that this is not so, and the recovery volume is independent of the duration of moderate steady-state exercise (25).

One practical difficulty has beset measurement of the recovery volume from the earliest investigations. As far back as 1913, it was known that the rate of oxygen uptake remained elevated for several hours after severe exercise (26). Hill and his colleagues (27) were also aware of this and commented: "This rise in the basal metabolism is not part of the recovery process, but an effect of the general circulatory and metabolic disturbance produced by exercise." The distinction drawn is, of course, a theoretical rather than an observational one, and it is the present thesis that one should regard the whole process underlying the recovery volume as the "metabolic disturbance" produced by exercise.

A number of experiments have been carried out to confirm the oxygen debt theory simply in terms of volumes of oxygen under conditions in which the supply of oxygen is restricted. Experiments, both in ischaemia (22, 28) and hypoxia (21, 29), have failed to confirm the theory in these simple terms. In rheumatic heart disease, where oxygen transport is limited and lactate production is increased, the relations between the rate of exercise on the one hand and the lag volume, "exercise rate", and recovery volume on the other hand do not differ from normal (30).

In the face of such experimental results it seems clear that the original theory does not hold true even in simple terms of volumes of oxygen.

For all these reasons it is proposed that the term oxygen debt should be discarded. It is based on an original theory which no serious physiologist or biochemist could now accept and the modifications which have been made to the original theory are so many and so complex that the meaning of the term oxygen debt is no longer uniform.

Although it is tempting to ascribe the recovery volume to the reconstitution of depleted high-energy-phosphate stores in muscle, the total magnitude of such stores does not seem nearly sufficient to account for the maximal magnitude of the recovery volume (30).

There must be other factors which determine the magnitude of the recovery volume of oxygen. Among these are possible alterations in mitochondrial P:O ratio, intra-cellular and trans-cellular movements of ions, and the general metabolic effects of an increase in body temperature and the release of catecholamines. How fruitful it is to examine the recovery volume in these terms is doubtful. In any biological system, a square-wave input gives rise to an output with a shape similar to that which we have been discussing. One thing, however, seems clear. The recovery volume is not determined in any causative way by the concentration of lactate in the blood after exercise.

References

1. FLETCHER, W.M., HOPKINS, F.G.: Lactic acid in amphibian muscle. J. Physiol. (Lond.) 35, 247-309 (1907)
2. HILL, A.V.: The heat produced in contracture and muscular tone. J. Physiol. (Lond.) 40, 389-403 (1910)
3. HILL, A.V.: The energy degraded in the recovery processes of stimulated muscles. J. Physiol. (Lond.) 46, 28-80 (1913)
4. HILL, A.V.: The oxidative removal of lactic acid. J. Physiol. (Lond.), 48, x-xix (1914)
5. MEYERHOF, O.: Die Energieumwandlungen im Muskel. 1. Über die Beziehungen der Milchsäure zur Wärmebildung und Arbeitsleistung des Muskels in der Anaerobiose. Pflueger's Arch. 182, 232-283 (1920)
6. MEYERHOF, O.: Über die Energieumwandlungen im Muskel. 2. Das Schicksal der Milchsäure in der Erholungsperiode des Muskels. Pflueger's Arch. 182, 284-317 (1920)
7. MEYERHOF, O.: Die Energieumwandlungen im Muskel. 3. Kohlenhydrat- und Milchsäureumsatz im Froschmuskel. Pflueger's Arch. 185, 11-32 (1920)
8. HILL, A.V., LUPTON, H.: Muscular exercise, lactic acid and the supply and utilisation of oxygen. Q. J. Med. 16, 135-171 (1923)

9. MARGARIA, R., EDWARDS, H.T., DILL, D.B.: The possible mechanisms of contracting and paying the oxygen debt and the role of lactic acid in muscular contraction. Am. J. Physiol. 106, 689-715 (1933)
10. BANG, O.: The lactate content of the blood during and after muscular exercise in man. Scand. Arch. Physiol. [Suppl. 10] 74, 49-82 (1936)
11. HARRIS, P., BATEMAN, M., BAYLEY, T.J., DONALD, K.W., GLOSTER, J., WHITEHEAD, T.: Observations on the course of the metabolic events accompanying mild exercise. Q. J. Exp. Physiol. 53, 43-64 (1968)
12. HUCKABEE, W.E.: Relationships of pyruvate and lactate during anaerobic metabolism. I. Effects of infusion of pyruvate or glucose and of hyperventilation. J. Clin. Invest. 37, 244-263 (1958)
13. HUCKABEE, W.E.: Relationships of pyruvate and lactate during anaerobic metabolism. II. Exercise and the formation of O_2-debt. J. Clin. Invest. 37, 255-263 (1958)
14. HARRIS, P., BATEMAN, M., GLOSTER, J.: Relations between the cardiorespiratory effects of exercise and the arterial concentration of lactate and pyruvate in patients with rheumatic heart disease. Clin. Sci. 23, 531-543 (1962)
15. HULTMAN, E.: Physiological role of muscle glycogen in man, with special reference to exercise. Circ. Res. 20-21 [Suppl. 1], 99-114 (1967)
16. DRURY, D.R., WICK, A.N.: Chemistry and metabolism of L (+) and D (=) lactate acids. Ann. N.Y. Acad. Sci. 119, 1061-1069 (1965)
17. OPIE, L.H., NEWSHOLME, E.A.: The activities of fructose 1,6-diphosphatase, phosphofructokinase and phosphoenolpyruvate carboxykinase in white muscle and red muscle. Biochem. J. 103, 391-399 (1967)
18. HILL, A.V., LONG, C.N.H., LUPTON, H.: Muscular exercise, lactic acid, and the supply and utilisation of oxygen. Parts VII to VIII. Proc. R. Soc. Lond. [Bid.] 97, 155-176 (1924)
19. KROGH, A., LINDHARD, J.: The changes in respiration at the transition from work to rest. J. Physiol. (Lond.), 53, 431-437 (1920)
20. LUPTON, H.: The recovery oxygen-usage after exercise in man. J. Physiol. (Lond.) 56, xvii-xix (1922)
21. SCHNEIDER, E.C., CLARKE, R.W.: Studies on muscular exercise under low barometric pressure. I. The consumption of oxygen and the oxygen debt. Am. J. Physiol. 74, 334-353 (1925)
22. ASMUSSEN, E.: Aerobic recovery after anaerobiosis in rest and work. Acta Physiol. Scand. 11, 197-210 (1946)
23. CHRISTENSEN, E.H., HÖGBERG, P.: Steady-state O_2-deficit and O_2-debt at severe work. Arbeitsphysiologie 14, 251-254 (195o)
24. DURAND, J., PANNIER, C.I., de LATTRE, J., MARTINEAUD, J.P., VERPILLAT, J.M.: The cost of the oxygen debt at high altitude. In: Exercise at altitude. Margaria, R. (ed.), pp. 40-47. Amsterdam: Excerpta Medica (1967)
25. SCHNEIDER, E.G., ROBINSON, S., NEWTON, J.: The oxygen dept in aerobic work. Physiologist 7, 247 (1964)
26. BENEDICT, F., CATHCART, E.P.: Muscular Work: a metabolic study with special reference to the efficiency of the human body as a machine. Washington: Carnegie Institute (1913)
27. HILL, A.V., LONG, C.N.H., LUPTON, H.: Muscular exercise, lactic acid, and the supply and utilisation of oxygen. Parts IV to VI. Proc. R. Soc. Lond. [Biol.] 97, 84-138 (1924)
28. BEER, G., YANCE, L.R.: Blood flow, oxygen uptake and capillary filtration in resting skeletal muscle. Am. J. Physiol. 223, 492-498 (1972)
29. ALPERT, N.R.: Lactate production and removal and the regulation of metabolism. Ann. N.Y. Acad. Sci. 119, 995-1011 (1965)
30. HARRIS, P.: Lactic acid and the phlogiston debt. Cardiovasc. Res., 3, 381-390 (1969)

Chapter 7

Oxygen Dept: Its Role and Significance

P. Cerretelli

Introduction

I am very grateful to the organisers of the Meeting for having invited me to participate in this debate on "oxygen debt" with Professor Harris and other distinguished colleagues. Since the meeting in Göteborg, in 1971, many statements have been made on the subject by a number of investigators. Even though their conclusions appear rather controversial, most of them are extremely useful in allowing some insight into the problem.

As a defender of the oxygen debt concept, I have the following aims in this presentation:

1) Critical revision of the classic definition of oxygen debt
2) Establishment of the role and the physiological significance of the oxygen debt, particularly of its "alactic" and "lactacid" fractions
3) Analysis of the fate of lactate during recovery from anaerobic exercise.

Alactic Oxygen Debt:

Definition, Origin, and Role

Upon a square wave of <u>submaximal</u> (<u>aerobic</u>) <u>work</u>, the V_{O_2} on-response curve on the whole organism, like that of an isolated perfused muscle, lags behind the homologous mechanical work tracing. Even when the adjustment of the latter is instantaneous, the half-time of the V_{O_2} on-response curve still ranges between 15 in the dog (whole animal (1) and isolated gastrocnemius (2) and 25-80 s in man for most experimental conditions (3, 4, 5). Correspondingly, an oxygen lag volume (6) is identified, i.e., an oxygen deficit equal to the area between the work on ordinate and the measured $\dot{V}_{O_2}$ is incurred. In the above experimental conditions the maximal measured debt for an adult man varies between 1.5 and 2.5 litres (7.5-12.5 kcal). Upon cessation of the activity, an oxygen recovery volume is found, classically referred to as the "oxygen debt payment" (5). This volume is determined as the time integral of the $\dot{V}_{O_2}$ off-response curve after correction for resting metabolism usually assumed to be equal to the pre-exercise level. The half-time of the latter curve in man is essentially constant at about 25-30 s, regardless of the kinetics of the $\dot{V}_{O_2}$ on-response. Correspondingly, the measured oxygen recovery volume may be found to be somewhat larger than (7, 8, 9), equal to (5), or even smaller than (3) the corresponding lag volume.

Various factors may be responsible for such differences.

1) When the <u>recovery volume is larger than the lag volume</u> (i.e., O_2 debt exceeds O_2 deficit):
a) The decreased efficiency of oxidative phosphorylation (~P:O ratio) due to the rise in muscle temperature following prolonged exercise (10) leads to a relative increase of the recovery-to-lag volume ratio
b) The amount of oxygen required by the respiratory and cardiac muscles during the early phase of recovery (computed as recovery volume)
c) More ~P may be hydrolysed than predicted from the oxygen lag volume, in which case part of the recovery volume would be used to restore this ~P
2) When the <u>recovery volume is smaller than the "lag" volume:</u> The situation where some lactate is produced during the transition from rest to steady-state oxygen consumption due to a transient anaerobic phase (3, 11, 12) is potentially characterised by a lag-to-recovery volume ratio higher than 1. In fact, the lag volume, besides the oxygen saved through the PC hydrolysis and the aliquot corresponding to the utilisation of the oxygen stores, includes the energy (O_2) equivalent of anaerobic glycolysis, the payment of which is not confined to the time when the recovery volume is measured.

A debt incurred without lactic acid formation, the most common occurrence in moderate rhythmic exercise such as walking, running, and cycling, is called alactacid (5) or alactic (8). The building up and payment of such a debt seem to be quantitatively related to hydrolysis and resynthesis, respectively, of some metabolite, most probably phosphocreatine. This hypothesis was first put forward by Margaria et al. (5), and has recently gained support from experiments carried out in the isolated perfused mammalian muscle (2, 13), where the time course of ~P resynthesis was found to be very close to that of the $\dot{V}_{O_2}$ off-response (Fig. 1) and to the time course in the human body (14). In addition, the ~P:O ratio determined in the isolated perfused dog gastrocnemius does not differ significantly from 3, i.e., the value resulting from the accepted stoichiometry of oxidative phosphorylation (13). Even though the above findings cannot be considered direct evidence, the experimental results are compatible with a cause-effect relationship. The oxygen obtained from the depletion of the oxygen stores could potentially be responsible for the time lag in the $\dot{V}_{O_2}$ adjustment, but it has been estimated in the isolated muscle at only 6% of the measured lag and recovery volumes (2). When this figure is extrapolated to the whole body the resulting amount (600-800 ml O_2; 15, 16) is again inadequate to account for the maximal oxygen volume missing at the onset of work (1.5-2.5 litres). No alternative exoergonic reaction that may be responsible for the building up of the alactic debt has so far been identified in the muscle.

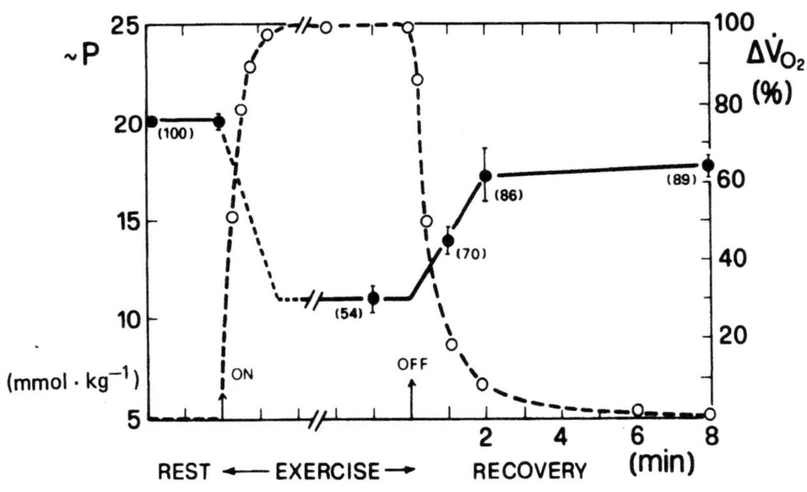

Fig. 1. $\dot{V}_{O_2}$ on- and off-response of an isolated perfused dog gastrocnemius upon and after stimulation with 20 short (2 s) tetani per minute (broken line). Piiper et al., 1968; Piiper and Spiller, 1970; and time course of the high-energy phosphate (~P) resynthesis during recovery following stimulation (solid line). Piiper and Spiller, 1970

The alactic mechanism plays a fundamental role in the adaptation to work, as (1) it is a prerequisite for control metabolites (ATP, ADP, AMP) to attain concentration levels suitable to remove inhibition of PFK and thus promote an adequate "energy status" of the muscle cell (17); (2) it allows for sudden changes of energy output, which are otherwise impossible due to the time lag of the adjustment of the oxidative machinery; and (3) it permits the development over short periods of time of power peaks three times as high as can be provided by oxidation (18).

Lactacid Oxygen Debt

A Revised Interpretation

When the load imposed on a muscle or on the body is supramaximal (i.e., more power is developed than can be supplied by maximal oxidation, $\dot{V}_{O_{2max}}$) and is sustained for at least a few minutes, the measured oxygen recovery volume always exceeds the corresponding oxygen lag volume (5, 19, 20). In the above conditions the energy for muscle contraction is made available to some extent by hydrolysis of phosphocreatine, as for aerobic work (see previous section), but mainly by anaerobic glycolysis, as revealed by lactate production. Margaria and his colleagues were the first to attempt the quantitative evaluation of the glycolytic energy yield (i.e., of the lactacid oxygen debt built up) as a consequence of a supramaximal load. Two different approaches were followed 30 years apart.

In the early nineteen-thirties, Margaria et al. (5) tried to estimate the amount of energy released by glycolysis from analysis of the $\dot{V}_{O_2}$ off-response and of the blood lactate recovery curves. The oxygen consumption curve during recovery was considered to result from the sum of four functions:

1) The basal oxygen consumption measured before the performance of work
2) An oxygen consumption attributable to oxidation of lactic acid, decreasing as an exponential function of time, with a half-time of about 15 min
3) Another exponential decrease, with a half-time of about 30 s
4) An oxygen consumption decreasing over several hours after cessation of exercise.

Of these four functions, (2) and (3) are the only ones that in the words of the authors: "have the meaning of an oxygen debt payment". Component (3), i.e., the "paid" alactic debt, is indicative of the homologous debt built up, with the reservations made in the Section "The Alactic Oxygen Debt". Function (2), i.e., the lactic acid mechanism described by Hill, is the lactacid oxygen debt paid, which was considered by the authors to be an indicator of the oxygen debt built up in proportion to the accumulated lactate. This is the aliquot of the recovery volume referred to in the literature simply as the lactacid oxygen debt. In some cases, and for a given set of conditions, the debt paid may be a constant fraction of the debt built up. However, taking one for the other is a conceptual mistake that has caused confusion and misunderstanding among investigators. In fact, the size of the paid debt does not allow any precise quantitative estimate of the amount of anaerobic metabolism, i.e., of the oxygen debt actually built up, unless: (1) the efficiency of the process of glycogen resynthesis is known; (2) oxidation of lactate as a fuel during recovery can be ruled out; and (3) a quantitative estimate of the amount of lactate produced can be obtained.

Component (2) of Margaria et al. (5), i.e., the lactacid oxygen debt paid includes:

1) The oxygen required for the resynthesis of glycogen from lactate
2) The energy wasted in the above transformation, i.e., about 60% of the total (the energy released by the formation of 1 mol lactate is about 20-25 kcal, while the cost of resynthesis of glycogen is 55 kcal (21)).

Fraction (2) does not correspond to a debt built up. Its computation as part of Factor (2) leads to overestimation of the actual size of the oxygen debt. By contrast, the oxygen equivalent of the lactate possibly oxidised as a fuel during recovery (sometimes a very high fraction of the total) is missed from the computation of the debt built up, as it obviously does not appear as part of component (2). The two above aliquots may partially compensate for each-other, and this explains why occasionally reasonable values for the oxygen debt built up could be found from the analysis of the debt paid, i.e., of component (2).

Thirty years later, when the problem of the measurement of the glycolytic energy release seemed to have fallen into oblivion, as most experimental work was aimed at the study of lactate catabolism during recovery, Margaria et al. (22) proposed estimation of the contribution of anaerobic glycolysis to the energy balance of supramaximal exercise from the actual lactate accumulated in the body, rather than from the oxygen debt paid. This approach required the notion of the amount of lactate actually stored in the tissues at the end of an exercise (not of the aliquot possibly metabolised during work, because this is already computed in the oxygen consumption) as well as the knowledge of its in vivo energy (or oxygen) equivalent. While a precise quantitation of the lactate concentration in the various tissues at the end of the exercise appeared to be a rather formidable task, an approximate estimate of the overall lactate accumulated in the body was attempted. The basic assumption was that the lactate produced by a given muscle group 6-8 min after discontinuation of work was redistributed uniformly in all intra- and extracellular fluids even in the muscles not directly involved in the exercise, as shown by Karlsson et al. (23), and could therefore be evaluated simply from blood lactate concentration and total body water. This assumption obviously did not imply that at equilibrium no concentration gradient for lactate was established across the cell membrane. In fact, the lactate concentration ratio of intracellular to extracellular compartments was shown by Roos (24) to be approximately 0.4 in rat diaphragm at equilibrium, at intra- and extracellular pH values of 7.0 and 7.4, respectively.

The in vivo oxygen equivalent of glycolysis was determined in the isolated perfused dog gastrocnemius by a stop-flow technique (25). The formation of 1 g lactate was found to be associated with the release of 248 calories, a value similar to the results obtained from in vitro experiments.

Thus, reliable estimates of the energy released by glycolysis were made in various experimental conditions and compared with the measured oxygen lag volumes. On the whole, despite the criticism raised against the procedure followed for the estimation of lactate, the two sets of measurements proved to correlate well in a variety of experimental conditions including acute and chronic hypoxia.

Regardless of the precision of the approach, it can be stated that in supramaximal work a certain amount of lactate is produced and that the energy yielded replaces missing oxygen, i.e., part of the lag volume. Thus, an oxygen debt is built up. It would be meaningless, however, to compare the extent of the lactacid oxygen debt built up calculated as in this Section with component (2) of the recovery curve, i.e., the debt paid. These, as indicated earlier, are in fact two different entities.

The Fate of Lactate During Recovery

Since Huckabee's (26, 27) unsuccessful (16, 28) attempt to correlate the payment of the oxygen debt with the excess lactate, i.e., to the part of the rise in lactate concentration that is left over after subtracting the aliquot due only to the increase in the concentration of pyruvate, a number of interesting studies dealing with the catabolism of lactate and the relationship between oxygen debt paid and other blood biochemical changes have appeared. Kayne and Alpert (29) showed a complete dissociation between oxygen consumption and metabolite removal during recovery from exercise in dog. In fact, hepatectomy did not significantly alter the pattern of recovery oxygen consumption, whereas that of metabolite removal was markedly changed. Minaire (30) showed that the fate of infused lactate in running dogs was to a large extent oxidative, while Depocas et al. (31) and Hubbard (32) also showed a large lactate turnover in aerobic exercise. Brooks et al. (33) concluded that in the rat the primary fate of lactic acid produced in anaerobic exercise was oxidative. On the other hand, Rowell et al. (34) indicated that during moderate prolonged exercise, a very significant fraction of the lactate produced (perhaps 50%) is resynthesised into glycogen by the liver, while Hermansen and Vaage (35) have recently come to the conclusion that in man 75% of the lactate found in the muscles at the end of an anaerobic exercise is converted to glycogen in the same muscle during recovery. Moreover, Gisolfi et al. (36) showed that the measured lactacid oxygen debt is reduced if during recovery following a supramaximal exercise the subject carries out an aerobic load, suggesting that a greater fraction of lactate may have been utilised as substrate. The same conclusion was also reached, indirectly, by Davies et al. (37), who showed in man that moderate work loads increased the rate of disappearance of lactate from blood during recovery.

From the above literature it may be concluded that the fate of lactate during recovery from anaerobic exercise is not necessarily glycogen resynthesis but, mainly, oxidation. Thus, as indicated before, the magnitude of the lactacid oxygen debt built up during the exercise cannot be inferred from the analysis of the oxygen recovery volume.

Figure 2 is a schematic plot of the glycolytic pathway for three limiting conditions, i.e., (1) that in which exercise is aerobic and, obviously, the rate of pyruvate oxidation equals that of production with no lactate accumulation (Fig. 2a); (2) a condition of supramaximal exercise in which a lactacid oxygen debt is built but not paid, since lactate is totally oxidised as a substrate in the course of recovery (Fig. 2b); and (3) a situation identical with (2) except for the recovery, when all lactate is converted to glycogen and a maximal oxygen debt is paid (Fig. 2c). Supramaximal exercise could theoretically be characterised by any pattern between those shown in Fig. 2b and c.

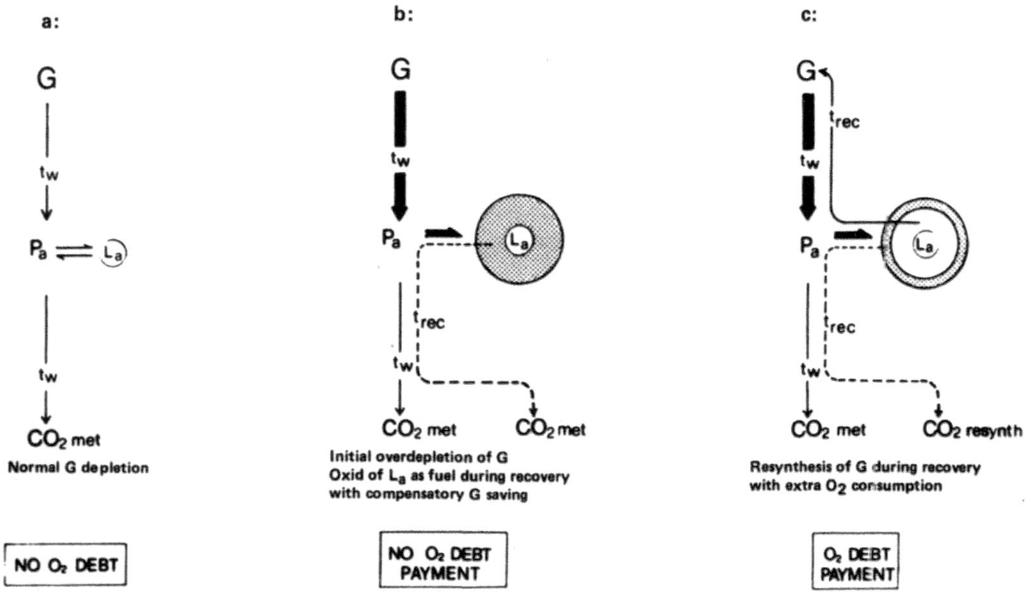

Fig. 2a–c. Schematic representation of the glycolytic pathway in three limiting conditions: steady-state aerobic exercise (a), supramaximal anaerobic exercise and subsequent recovery characterised by total oxidation of lactate (b), and supramaximal anaerobic exercise and recovery characterised by glycogen resynthesis from lactate (c). Outer shell depicts theoretical lactate energy equivalent required for resynthesis of glycogen (G) from the inner shell; t_w, working time; t_{rec}, recovery phase; Pa, pyruvate concentration; La, lactate concentration; CO_2 met, metabolic CO_2; CO_2 resynt, CO_2 produced for glycogen resynthesis

The processing of 1 mol lactate (about the maximum that can be accumulated in the human body at the end of strenuous exercise) through oxidation (Fig. 2b) implies the consumption of 3 mol O_2 (67 litres), whereas glycogen resynthesis (Fig. 2c) requires 0.5 mol (11 litres) (21). The fate of the energy deriving from these two routes, however, is different. The energy released by oxidation is made available just to the body and may therefore replace energy derived from other substrates, whereas the oxygen taken up for glycogen resynthesis represents an extra metabolic load, since it is used for glycogen resynthesis. The latter appears in the recovery as component (2) of Margaria et al. (5).

In practice, upon accumulation of one mol lactate, the oxygen uptake in the course of recovery before the pre-exercise blood lactate level is attained (60-75 min or more) may be 20-25 litres, i.e., from about half to one-third of that potentially required to burn up the total lactate previously accumulated. This finding rules out oxi-

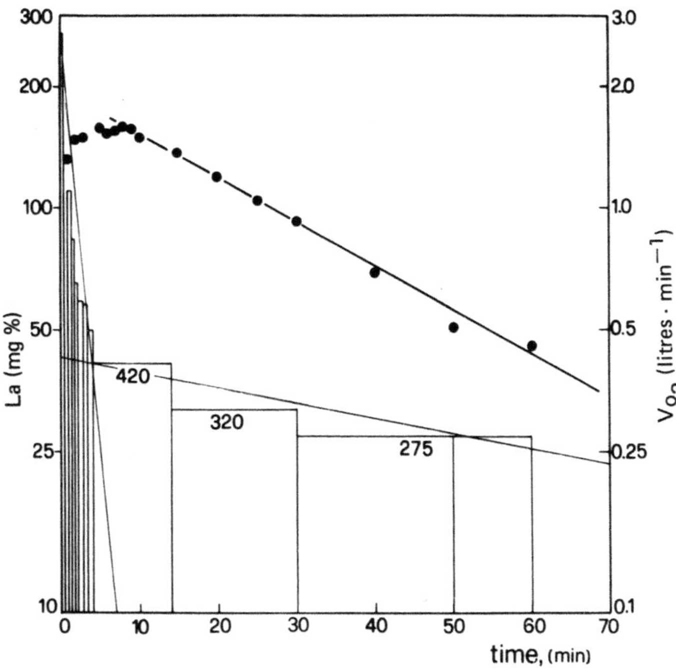

Fig. 3. Lactic acid disappearance curve and total oxygen uptake during recovery after an all-out swimming performance. The calculated O_2 recovery volume within the first 60 min of recovery is 23.5 litres, out of which 2.6 litres are the alactic fraction of the debt and 2.9 litres the sum of components (2) and (4) of Margaria et al. (27). The amount of lactate metabolised over the same period may be estimated at 550 mmol: its oxidation would require approximately 37 litres O_2

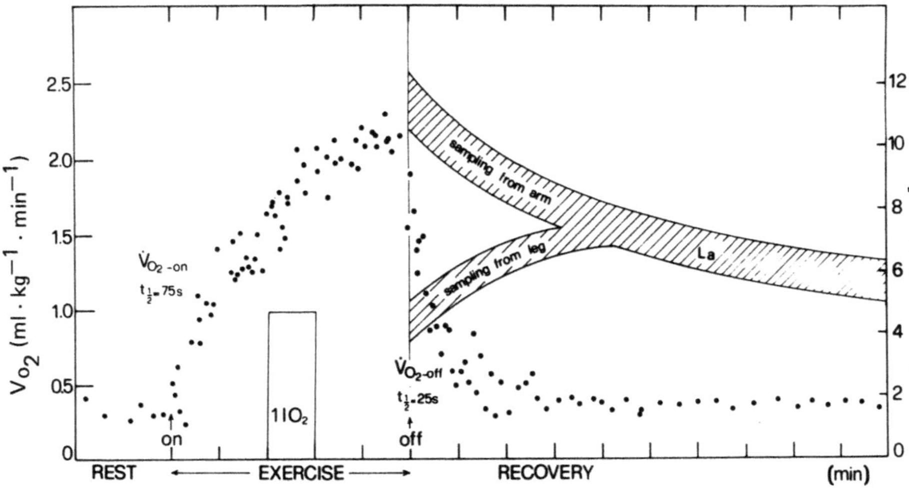

Fig. 4. Breath-by-breath O_2 uptake at the onset of arm cranking (125 W; supine position) and during the subsequent recovery phase. Blood lactate recovery curves from leg and arm samples are also drawn. The calculated O_2 volume required to oxidise the 315 mmol lactate produced would be 23 litres, approximately the total amount taken up by the subject during the 60-min recovery

dation during the recovery period as the exclusive fate of lactate. The calculated component (2) of Margaria et al. (5) may vary widely in size, depending on the conditions of the experiment. After treadmill running it is usually large (5-6 litres), indicating that during recovery some resynthesis of glycogen has taken place. By contrast, after swimming (Fig. 3) and arm cranking (Fig. 4), in spite of the relatively large lactate accumulation, component (2) is smaller and can sometimes hardly be identified. After subtraction of the alactic fraction, the overall oxygen recovery volumes, i.e., components (2) and (4) of Margaria et al. (5), according to recent measurements (carried out in collaboration with Drs. David Pendergast and D.W. Rennie of SUNY at Buffalo) may amount to 2.9 and 0.5 litres in the above conditions, with initial lactate accumulations of 550 and 315 mmol, respectively. The oxygen necessary to oxidise lactate may be estimated at 37 and 23 litres, respectively, for the two conditions (for calculation, see legends to Figs. 3 and 4). In the former case (swimming) a large fraction of the lactate produced was evidently oxidised and component (2) associated with resynthesis may have become too small to be resolved by conventional analytical procedures. During arm cranking the energy released by lactate oxidation may have been very close to the amount utilised by the body during recovery, and glycogen resynthesis may have been practically nil (Fig. 2b).

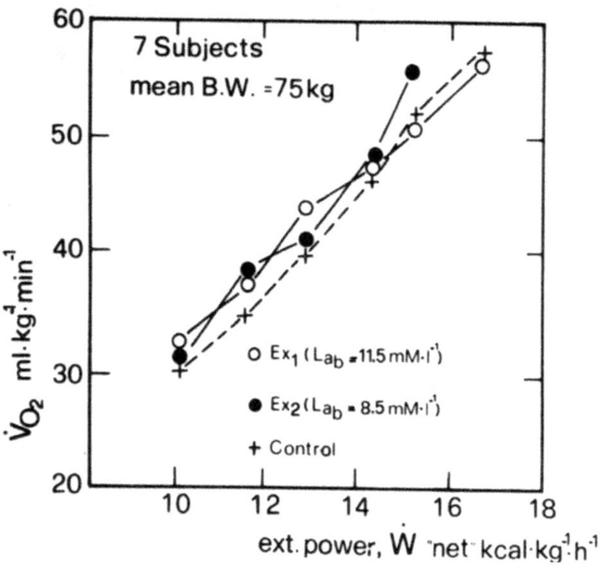

Fig. 5. $\dot{V}_{O_2}$ as a function of external (mechanical) power during treadmill running by seven subjects with mean body weight of 75 kg. Ex_1 (o) and Ex_2 (•) refer to conditions in which graded loads (2 min each) were carried out in the course of recovery following preceding anaerobic exercises causing blood lactate levels of 11.5 mmol·litre^{-1} and 8.5 mmol·litre^{-1}, respectively. Control levels (+) are also shown

In Fig. 5 (38), the energy cost (ml·kg^{-1}·min^{-1}) of graded submaximal running exercises performed during recovery following two priming anaerobic loads is compared with control values. No significant $\dot{V}_{O_2}$ differences could be detected at any of the loads. Thus, no extra oxygen cost was added to control levels, as would have been required by a simultaneous oxygen debt payment associated with glycogen resynthesis. Assuming a debt payment of 6 litres, the cost of running would have been expected to be at least 10% higher in the test than in control conditions. The results of this experiment are compatible with those of Gisolfi et al. (36) and with the situation depicted in Fig. 2b.

From the above data it may be concluded that the fate of lactate formed during anaerobic exercise seems to be preferentially oxidative, the prerequisite being the capability of the body to utilise the energy derived from the process immediately in the course of recovery. If the amount of lactate accumulated is large and recovery is carried out at rest, the aliquot of lactate in excess to that oxidised appears to be resynthesised and an oxygen debt is paid. Conversely, if the amount of lactate is small or if the body metabolic load in the course of recovery is high (active recovery), lactate is used as the most readily available substrate and no oxygen debt payment can be found.

Acknowledgement

The author is very much indebted to Professor R. Margaria for his friendly and helpful criticism of the paper.
Part of this work was completed during the author's tenure of the Dr. Henry C. and Bertha H. Buswell Research Fellowship at the State University of New York at Buffalo (USA).

Summary

At the onset of aerobic exercise the $\dot{V}_{O_2}$ on-response of the whole body lags behind the adjustment of the mechanical power output that can be made practically instantaneously.

Correspondingly, an alactic oxygen debt is built up (lag volume), which is causally related to hydrolysis of the high-energy phosphates (~P), mostly phosphocreatine.

After discontinuation of work the alactic debt is paid, i.e., some oxygen is used (recovery volume) to replenish the ~P stores.

In the course of supramaximal exercise a sizeable amount of energy is released by anaerobic glycolysis and a lactacid oxygen debt is built up.

During recovery following supramaximal exercise the accumulated lactate can be either oxidised or resynthesised into glycogen.

The oxygen debt paid is indicative only of the fraction of re-synthesised glycogen. If lactate is fully oxidised due to increased metabolism, no oxygen debt appears to be paid.

Since the fraction of oxidised lactate varies with the experimental conditions in recovery, no precise estimate of the oxygen debt built up can be made from the amount of debt paid.

The lactacid oxygen debt built up can be evaluated only if the amount of lactate accumulation in the body and its energy (oxygen) equivalent are known.

References

1. MARCONI, C., PENDERGAST, D., KRASNEY, J., CERRETELLI, P.: Kinetics of the $\dot{V}_{O_2}$ on -and off- responses in the running dog. Physiologist 21: 76 (1978)
2. PIIPER, J., di PRAMPERO, P.E., CERRETELLI, P.: Oxygen debt and high energy phosphates in gastrocnemius muscle of the dog. Am. J. Physiol. 215, 523-531 (1968)
3. CERRETELLI, P., SHINDELL, D., PENDERGAST, D.P., di PRAMPERO, P.E., RENNIE, D.W.: Oxygen uptake transients at the onset and offset of arm and leg work. Respir. Physiol. 30, 81-97 (1977)
4. LINNARSSON, D.: Dynamics of pulmonary gas exchange and heart rate changes at start and end of exercise. Acta Physiol. Scand. [Suppl.] 415, 1-68 (1974)
5. MARGARIA, R., EDWARDS, H.T., DILL, D.B.: The possible mechanism of contracting and paying the oxygen debt and the role of lactic acid in muscular contraction. Am. J. Physiol. 106, 689-715 (1933)
6. KROGH, A., LINDHARD, J.: The changes in respiration at the transient from work to rest. J. Physiol. 53, 431-437 (1920)
7. CHRISTENSEN, E.H., HÖGBERG, P.: Steady state, O_2 deficit and O_2 debt at severe work. Arbeitsphysiologie 14, 251-254 (1950)
8. HENRY, F.M.: Aerobic oxygen consumption and alactic debt in muscular work. J. Appl. Physiol. 3, 427-438 (1951)
9. LUKIN, L., RALSTON, H.J.: Oxygen deficit and repayment in exercise. Intern. Z. Angew. Physiol. 19, 183-193 (1962)
10. BROOKS, G.A., HITTELMAN, K.J., FAULKNER, J.A., BEYER, R.E.: Temperature, skeletal muscle mitochondrial functions, and oxygen debt. Am. J. Physiol. 220, 1053-1066 (1971)
11. BANG, O.: The lactate content of the blood during and after muscular exercise in man. Scand. Arch. Physiol. 74, 49-82 (1936)
12. SAIKI, H., MARGARIA, R., CUTTICA, F.: Lactic acid production in submaximal work. Intern. Z. Angew. Physiol. 24, 57-61 (1967)
13. PIIPER, J., SPILLER, P.: Repayment of O_2 debt and resynthesis of high energy phosphates in gastrocnemius muscle of the dog. J. Appl. Physiol. 28, 657-662 (1970)
14. HULTMAN, E., BERGSTROM, J., McLENNAN, N.: Anderson, Breakdown and resynthesis of phosphorylcreatine and adenosine triphosphate in connection with muscular work in man. Scand. J. Clin. Lab. Invest. 19, 56-66 (1967)
15. Di PRAMPERO, P.E.: Energétique de l'exercice musculaire. J. Physiol. (Paris) 65, 51A-86A (1972)
16. HARRIS, P.: Lactic acid and the phlogiston debt. Cardiovasc. Res. 3, 381-390 (1969)
17. NEWSHOLME, E.A., START, C.: Regulation in Metabolism. Pp. 106-117. London: Wiley 1974
18. MARGARIA, R., CERRETELLI, P., MANGILI, F.: Balance and kinetics of anaerobic energy release during strenuous exercise in man. J. Appl. Physiol. 19, 623-628 (1964)
19. HILL, A.V., LONG, C.N.H., LUPTON, H.: Muscular exercise, lactic acid, and the supply and utilization of oxygen. Parts VII to VIII. Proc. R. Soc. B, 97, 155-176 (1924)
20. KNUTTGEN, H.G.: Oxygen debt after submaximal physical exercise. J. Appl. Physiol. 29, 651-657 (1970)
21. McGILVERY, R.W.: "Biochemistry - A functional approach", pp. 280-282. London, Philadelphia: W.B. Saunders 1970

22. MARGARIA, R., CERRETELLI, P., di PRAMPERO, P.E., MASSARI, C., TORELLI, G.: Kinetics and mechanism of oxygen debt contraction in man. J. Appl. Physiol. 18, 371-377 (1963)
23. KARLSSON, J., BONDE PETERSEN, F., HENRIKSSON, J., KNUTTGEN, H.G.: Effects of previous exercise with arms or legs on metabolism and performance in exhaustive exercise. J. Appl. Physiol. 38, 763-767 (1975)
24. ROOS, A.: Intracellular pH and distribution of weak acids across cell membrane. A study of D- and L- lactate and of DMO in rat diaphragm. J. Physiol. (Lond.) 249 1-15 (1975)
25. CERRETELLI, P., di PRAMPERO, P.E., PIIPER, J.: Energy balance of anaerobic work in dog gastrocnemius muscle. Am. J. Physiol. 217, 581-585 (1969)
26. HUCKABEE, W.E.: Relationships of pyruvate and lactate during anaerobic metabolism. I-Effect of infusion of pyruvate or glucose and of hyperventilation. J. Clin. Invest. 37, 144-254 (1958)
27. HUCKABEE, W.E.: Relationships of pyruvate and lactate during anaerobic metabolism. II-Exercise and formation of O_2 debt. J. Clin. Invest. 37, 255-263 (1958)
28. WASSERMAN, K., BURTON, G.G., van KESSEL, A.L.: Excess lactate concept and oxygen debt of exercise. J. Appl. Physiol. 20, 1299-1306 (1965)
29. KAYNE, H.L., ALPERT, N.R.: Oxygen consumption following exercise in the anesthetized dog. Am. J. Physiol. 206, 51-56 (1964)
30. MINAIRE, Y.: Possibilités d'utilisation musculaire de l'acide lactique lors du frisson et de l'exercise. Lyon: Emmanuel Vitte 1965
31. DEPOCAS, F., MINAIRE, Y., CHATONNET, J.: Rates of formation and oxidation of lactic acid in dogs at rest and during moderate exercise. Can. J. Physiol. Pharmacol. 47, 603-610 (1969)
32. HUBBARD, J.: The effect of exercise on lactate metabolism. J. Physiol. 231, 1-18 (1973)
33. BROOKS, G.A., BRAUNER, K.E., CASSENS, R.G.: Glycogen synthesis and metabolism of lactic acid after exercise. Am. J. Physiol. 224, 1162-1166 (1973)
34. ROWELL, L.B., KRANING II, K.K., EVANS, T.O., KENNEDY, J.W., BLACKMON, J.R., KUSUMI, F.: Splanchnic removal of lactate and pyruvate during prolonged exercise in man. J. Appl. Physiol. 21, 1773-1783 (1966)
35. HERMANSEN, L., VAAGE, Q.: Lactate disappearance and glycogen synthesis in human muscle after maximal exercise. Am. J. Physiol. 233, E422-E429 (1977)
36. GISOLFI, F., ROBINSON, S., TURRELL, E.S.: Effects of aerobic work performed during recovery from exhausting work. J. Appl. Physiol. 21, 1767-1772 (1966)
37. DAVIES, C.T.M., KNIBBS, A.V., MUSGROVE, J.: The effect of recovery exercise on the removal of lactic acid from the blood. J. Physiol. 196, 61 (1968)
38. LEIBOWITZ, R.: The effect of pre-exhaustive treadmill exercise on maximal aerobic power in college age males. Masters Thesis, State University of New York at Buffalo 1977

Discussion

The challenging title of Dr. Harris's (1) paper: '"Oxygen debt" does not exist', and Professor Cerretelli's (2) spirited description of himself as "a defender of the oxygen debt concept" suggest that our two speakers disagree with each other to a much greater extent than, in my opinion, is the case. Neither speaker, and for that matter no-one else in the room, doubts for an instant that the phenomenon that led Hill (3) to coin the phrase "oxygen debt" does exist. At the onset of moderate aerobic exercise, the rate of oxygen absorption, $\dot{V}_{O_2}$, takes an appreciable time to rise from its resting value by an amount proportional to the intensity of exercise. Similarly, after the end of exercise the increased $\dot{V}_{O_2}$ takes an appreciable time to return to its resting level. This situation is admirably illustrated by Fig. 1 in Dr. Harris's article, and the facts are not in dispute.

Nor is the "concept" of oxygen debt really under attack, since the concept involves only a slight extrapolation from the above observations, which have been fully confirmed by recent direct biochemical study. Obviously in the early stages of moderate exercise, physical energy (work and heat) must be coming from chemical reactions other than the contemporaneous oxidation of substrates such as glycogen and fatty acids. We all even agree that the anaerobic energy sources are the hydrolysis of phosphocreatine and the formation of lactate from glycogen. It is also clear conceptually - but not in full detail experimentally - that extra oxygen must be consumed to return the various substrates to their resting state after exercise is finished.

The questions about which it appears there is still disagreement seem to me to be of two totally different types:

1) Is the terminology used to describe this phenomenon appropriate or is it misleading?
2) Do we understand the physiological and biochemical basis of what is observed?

1. Terminology

I do not regard this as a mere question of carping semantics. Giving something a name usually makes people feel that it has an identity, which can be a mistake. Giving it several different names implies falsely that it has several different identities, which is almost certainly a mistake. Certainly the variety of names and expressions employed is wide, and the possibility for confusion is correspondingly so. For the effect at the onset of exercise I have found "O_2 deficit" (2), "lag volume" (1, 2), "on-response" (2), "incurring an O_2 debt (4), "contraction of O_2 debt" (5). For the effect early in recovery one has the choice of "oxygen debt" (2), "recovery volume" (1), "off-response" (2) and "oxygen debt payment" (4, 5).

At the Meeting the question of an agreed terminology was discussed, and Professor Denolin's strongly expressed view was upheld - we should all stick to the names "deficit" and "debt". Though at the time this seemed innocuous enough I must confess to later doubts. The financial analogy is striking, but one is led to ask exactly what is supposed to be borrowing oxygen from what, and when? An analogy, to be useful, must be simpler than the thing it is trying to explain; and if we are obliged to hypothesise interest payments, depreciation, numbered Swiss bank accounts, or Cayman Island tax havens, we may well be on the wrong track.

The human body can increase its oxygen income (in contrast to its financial one) rapidly, but not instantly, at least 15-fold. The element of oxygen debt thus arises at the beginning of exercise, from the time lag in increasing the oxygen income. To me the terminology adopted at the Meeting obscures this fact. I think we would be wise to follow the pioneers Hill (4) and Margaria (5) in simply saying that an oxygen debt is "incurred" at the beginning of exercise; Margaria's expression "contracted" is correct English, but the word has other connotations, expecially in muscle. The oxygen debt is then "paid off" during recovery. This terminology locates the oxygen debt firmly in the onset period, not where it was located by the decision of our Symposium.

Wherever in time the debt is located, there remains the question as to what has borrowed oxygen from what. Clearly the muscles have not borrowed oxygen from the rest of the body, save to the very minor extent that they can do so by local diminution of their oxygen tension: even less has the body borrowed oxygen from the atmosphere. Things are actually the other way round: the oxygen debt is in fact a debt of negative oxygen. At this point we may well ask ourselves whether Dr. Harris is not right after all and admit that the phrase "oxygen debt" is indeed a very confusing one!

Surely what has really happened is that the contractile proteins have contracted an energy debt by dipping into the local stores of chemical free energy provided by PCr hydrolysis and lactate formation, rather than paying in full from the long-term income provided by oxidative phosphorylation. This is a debt indeed, and must ultimately be rapaid.

2. Physiological and Biochemical Mechanisms Involved

I shall be brief, because others at the Symposium are far more expert than I am. My impression is that we have a good understanding of the way in which oxygen consumption lags behind need at the onset of exercise, and of the way in which phosphocreatine is rebuilt during recovery. A much greater question mark, at least on the quantitative side, hangs over the fate of the considerable quantities of lactate that are often produced. Both Dr. Cerritelli (2) and Dr. Hermansen (6) have shown conclusively that it cannot possibly all be oxidised. Dr. Hermansen has rendered a great service by showing that lactate can be resynthesised into glycogen in human muscle as well as in the liver, thus coming full circle back to the point of view of Meyerhof. However, it does remain necessary to show that resynthesis of glycogen can account quantitatively for that part of the disappearance of lactate that is not covered by oxygen consumption. It was for lack of such quantitative comparisons in the "lactic acid era" that misleading comparisons came to be made between the oxidative recovery heat, lasting 10-20 min, and the rebuilding of glycogen, lasting more than 2o h (1).

Of course, it is easy for us to be wise after the event, but it does seem a pity that for so many years the rebuilding of glycogen in muscle was held to be impossible because of the supposed lack of certain key enzymes. I am reminded of one of my favourite quotations, of a question asked by Professor Mildred Cohen in discussion at another meeting (7): "You have certainly proved that it can't happen: but does it?"

D.R. Wilkie

References

1. HARRIS, P.: This volume, pp. 67-72.
2. CERRITELLI, P.: This volume, pp. 73-86.
3. HILL, A.V.: J. Physiol. 56, XXXII-XXXIII (1922)
4. HILL, A.V.: Trails and Trials in Physiology, p. 210, Edward Arnold: London (1965)
5. MARGARIA, R.: Biomechanics and energetics of muscular exercise. Fig. 1.27, p. 51. Clarendon: Oxford (1976)
6. HERMANSEN, L.: This volume pp. 46-66.
7. CIBA SYMPOSIUM 31. Energy Transformation in biological systems. Elsevier: Holland 1975

Chapter 8

Is the Use of Isotopic Tracers Valid for the Study of Lactate Metabolism in vivo

A. Freminet, C. Poyart, and L. Leclerc

Introduction

The use of isotopic tracers has been recommended by several authors in the past 10 years for the study of lactate metabolism in mammalian species (1). It is relevant to this Symposium devoted to anaerobic metabolism and lactic acid (1) to specify the significance of some parameters used in these studies, (2) to summarise the results obtained with these techniques, and (3) to comment on some practical or theoretical problems arising from this methodology.

Continuous Infusion Technique[1]

Methodology

In steady-state conditions, i.e., when the blood lactate concentration ([L] : $\mu mol.ml^{-1}$) does not vary over a certain period of time, the rate of lactate production (RaL) is equal to its rate of utilisation (RdL) and corresponds to the rate of lactate turnover (RL). This rate may be estimated by infusing lactate ^{14}C-U IV at a constant rate (F: $nCi.min^{-1}$ or $dpm.min^{-1}$) over a period of time sufficient for a steady level of lactate ^{14}C activity to be reached in the blood (A_L : $nCi.ml^{-1}$). This leads to a steady blood lactate specific activity (LSA : $nCi.\mu mol^{-1}$ = AL/[L]). In these conditions

$$RaL = RdL = RL\ (\mu mol.min^{-1}) = F/LSA \quad (1)$$

In non-steady-state conditions, RaL and/or RDL can be estimated from other mathematical expressions (2).

1 This discussion will not refer to the single injection technique

To determine the different sources of substrates contributing to RaL, the use of other tracers (e.g., glucose ^{14}C) is necessary. Lactate disposal in vivo is mainly due to liver gluconeogenesis and to oxidation to CO_2 and water in all tissues. It may appear easy to estimate these two pathways in vivo by measuring the ^{14}C incorporation into glucose molecules appearing in blood or into CO_2 molecules appearing in expired gases. These measurements allow calculation of two parameters:

1) The fraction of glucose derived from lactate = GSA/LSA x 2 (Eq. 2), where GSA is the glucose specific activity derived from infusion of lactate ^{14}C ($nCi.\mu mol^{-1}$). If the glucose production is known from previous determination of the glucose turnover, the rate of lactate conversion to glucose can be computed. Some authors have used more sophisticated expressions (3).

2) The fraction of lactate turnover oxidized = A $^{14}CO_2/F$ (Eq. 3) where A $^{14}CO_2$ is $^{14}CO_2$ activity in expired gases ($nCi.min^{-1}$). The rate of lactate oxidation is obtained from the product of this fraction and the R_L value.

Major Results

When expressed on a per $kg^{0.75}$ basis, the mean resting R_L has amounted to 40-50 $\mu mol.min^{-1}$ per $kg^{0.75}$ in the several mammalian species so far examined. This value corresponds to 1.0-1.2 $mmol.min^{-1}$ for a 70-kg human subject whose blood lactate level is 1 $\mu mol.ml^{-1}$ (or 1 $mmol.litre^{-1}$). In this condition, more than 50% of R_L was oxidized to CO_2 (corresponding to 15% of the total CO_2 production: $\dot{V}CO_2$) and 20%-30% was directed toward gluconeogenesis (1, 3).

In hyperlactataemic conditions such as are encountered during moderate hypoxaemia or produced by exogenous lactate load in normoxic conditions, R_L increased as [L] increased. The rate of lactate oxidation increased, representing 50% of the total energy expenditure of the resting organism when lactataemia was 6 $mmol.litre^{-1}$ (4).

The same phenomenon was observed following severe haemorrhage, burn injury, or muscular exercise. The participation of lactate in glucose synthesis was also increased and may have contributed to about half the glucose production (5-8).

Hyperlactataemia resulting from alcohol ingestion was shown to be related to an imbalance between RaL and RdL: RdL was decreased relative to RaL, mainly due to an inhibition of gluconeogenesis from lactate (2).

Hyperlactataemia following phenformin therapy has been related to an increase of both RaL and RdL, but the elevation of RaL was more pronounced than the increase of RdL (2).

In summary, lactate production (RaL) seems to be the main factor leading to hyperlactataemia, and if cellular respiratory and gluconeogenic processes are functional, lactate should be considered an important fuel for energy metabolism and a significant glucose precursor.

Criticisms of These Measurements

The rate of lactate turnover determined with the continuous lactate ^{14}C-U infusion technique is underestimated, as lactate ^{14}C is a reversible tracer. The presence of lactate ^{14}C in the blood does not only result from the dilution of F, but also from the lactate ^{14}C coming from newly synthesised substrates with the infused lactate ^{14}C (as for example, the sequence: lactate ^{14}C → glucose ^{14}C → lactate ^{14}C). In Equation (1), LSA is thus the sum of these two inputs: perfusion and recycling. Thus LSA is overestimated and R_L underestimated if the recycling of lactate ^{14}C is not taken into account.

The use of Equation (2) and (3) involves practical and theoretical problems

One practical problem is the difficulty of obtaining steady A $^{14}CO_2$ and steady GSA during the short-term experiments usually performed. For a precise estimation of A $^{14}CO_2$ the dilution effect of $^{14}CO_2$ due to the large CO_2 stores of the whole organism should also be taken into account (4).

The main theoretical problem is that with metabolites known to be either oxidized or used as gluconeogenic precursors, the use of [^{14}C-U] tracers may lead to wrong estimations of both oxidation and gluconeogenesis rates. It is possible, for instance, to observe ^{14}C incorporation into glucose without net glucose synthesis (during respiration) or to observe the appearance of $^{14}CO_2$ in expired gases during gluconeogenesis. As shown in Fig. 1, this problem arises form the fact that oxaloacetate (OAA) is a common metabolite for both gluconeogenesis and respiration pathways in gluconeogenic tissues. In these tissues, pyruvate ^{14}C-U derived from lactate ^{14}C-U may follow two different pathways:

Respiration. After the decarboxylation step ($^{14}CO_2$ formation), tagged acetyl CoA is condensed on untagged OAA (OAA type ⓐ) to form ci-

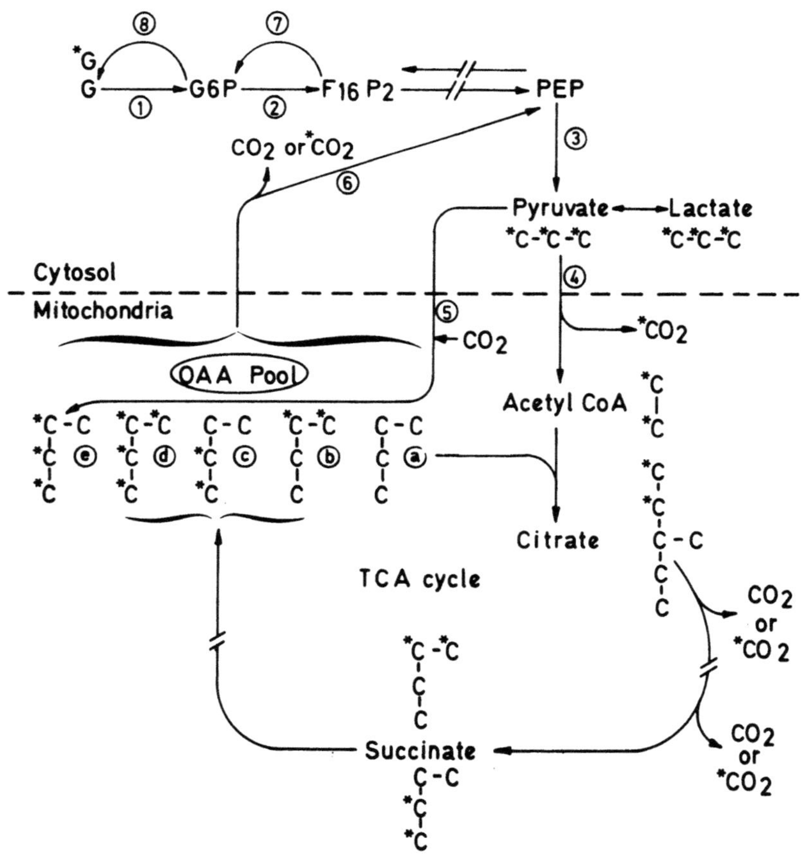

Fig. 1. Lactate metabolism in a mammalian organism following infusion of lactate ^{14}C-U. ★ indicates the tag

trate. In the course of the Krebs cycle, two CO_2 are produced (not tagged) and succinate ^{14}C is formed. As the succinate molecule is symetric al, two "types" of OAA molecules are present after this first run (types ⓑ and ⓒ). These OAA molecules again enter the tricarboxylic acid cycle with other tagged acetyl CoA molecules, and the CO_2 produced is either $^{14}CO_2$ (if citrate is made from OAA type ⓒ) or CO_2 (if citrate is made from OAA type ⓑ . OAA type ⓓ is produced after this second run of the cycle. In the continuation of the process any type of OAA (b→d) present may act as a $^{14}CO_2$ precursor.

Gluconeogenesis. The reaction catalysed by pyruvate kinase (3) is irreversible, and two steps have been described to re-form phosphoenolpyruvate (PEP): the carboxylation of pyruvate to OAA in the mitochondria (pyruvate carboxylase 5) and, after the indirect transfer of OAA to the cytosol, the formation of PEP from OAA (cytosolic phosphoenolpyruvate carboxykinase: PEPK 6). PEP is driven to glucose by the reversion of the glycolytic pathway due to fructose-1, 6-diphosphatase (7) and glucose-6-phosphatase (8).

When lactate ^{14}C is used as precursor, synthesised OAA will be labelled (type ⓔ). This form equilibrates with the common OAA pool (types a→d), so that when transferred to the cytosol any type of OAA ^{14}C may be picked up, eventually to form $^{14}CO_2$ during gluconeogenesis. At the same time, OAA type ⓔ may be included in the tricarboxylic acid cycle. It should be mentioned also that with a substrate ^{14}C passing only through the tricarboxylic cycle, some radioactivity may be recovered in the glucose molecules due to the exchange between OAA coming from carboxylation of pyruvate and the presence of OAA type ⓐ to ⓓ in the mitochondria. Conversely, a ^{14}C-labelled substrate only involved in glucose synthesis may lead to $^{14}CO_2$ formation by the equilibration between OAA type ⓔ and OAA used in the tricarboxylic acid cycle. Krebs and co-workers (9) have pointed out these difficulties arising from the use of ^{14}C tracers in biochemical physiology, in particular in the comments on results obtained with kidney slices (gluconeogenic tissues). These authors have proposed the term of "crossing-over" of carbon atoms between gluconeogenesis and respiration to describe this phenomenon.

One should note that this mechanism (Fig. 1) was described in rat gluconeogenic tissues, where PEPCK is essentially present in the cytosol (95%). In other species, PEPCK is present almost exclusively in the mitochondria (rabbit) or is present in equal quantities in the cytosol and in the mitochondria (man, guinea-pig). The results may differ depending upon the species under study (10).

To what extent these important criticisms of the use of ^{14}C tracers for the estimation of lactate metabolism in vivo render the results obtained by several authors invalid remains to be determined. It is our opinion that the definitive answer might not be found from in vivo or in vitro experiments.

References

1. MINAIRE, Y.: Origine et destinée du lactate plasmatique. J. Physiol. (Paris) 66, 229-257 (1973)
2. KREISBERG, R.: Glucose-lactate inter-relations in man. N. Engl. J. Med. 287, 132-137 (1976)
3. FREMINET, A., POYART, C.: Lactate-glucose interrelations, glucose recycling and the Cori-cycle in normal fed rats. Pfluegers Arch. 316, 25-31 (1975)
4. FREMINET, A., BURSAUX, E., POYART, C.: Effect of elevated lactataemia on the rates of lactate turnover and oxidation in rats. Pfluegers Arch. 346, 75-86 (1974)
5. WIENER, R., SPITZER, J.J.: Lactate metabolism following severe hemorrhage in the conscious dog. Am. J. Physiol. 227, 58-62 (1974)
6. WOLFE, R.R., MILLER, H.I., SPITZER, J.J.: Glucose and lactate kinetics in burn shock. Am. J. Physiol. 232, E415-E418 (1977)
7. ISSEKUTZ, B., SHAW, W.S., ISSEKUTZ, A.C.: Lactate metabolism in resting and exercising dogs. J. Appl. Physiol. 40, 312-319 (1976)
8. ELDRIDGE, F.L., T'SO, L., CHANG, H.: Relationships between turnover rate and blood concentration of lactate in normal dogs. J. Appl. Physiol. 37, 316-320 (1974)
9. KREBS, H.A., HEMS, R., WEIDEMANN, M.J., SPEAKE, R.N.: The fate of isotopic cerbon in kidney cortex synthesizing glucose from lactate. Biochem. J. 101, 242-249 (1966)
10. HANSON, P.W.: The choice of animal species for studies of metabolic regulation. Nutr. Rev. 32, 1-8 (1974)

Section II
Methodology

Opening Remarks

C.F. Rossel

During his investigations the physician or the researcher is confronted with the manifold problems relating to the experimentation. Bergson and Claude Bernard, in their philosophical analysis of the experiment, outline the theoretical and practical approaches: the experimental truths are objective but it is intuition or feeling that generates the experimental idea. The first stage, therefore, is a thorough analysis of the bases of the experimentation after objective observation of the natural phenomena. Now this objective observation - or the analysis of the bases of the experimentation - is beset with difficulties, especially when the human mind no longer grasps the object directly. This is particularly true when the physician has to resort to physicochemical methods to demonstrate a metabolic substance. He must then devote a great deal of attention to keeping his investigation method constant, naturally after analysing it thoroughly. Too often in clinical work the physician neglects this step in the experimentation, and leaves the choice of method to the chemist alone.

I therefore wish to thank the organisers of this symposium very warmly for their brilliant idea of devoting an entire session to the analysis of methods for the determination of lactic acid, in such a way as to raise certain problems:

1) Whether the so-called enzymatic methods for the analysis of lactic acid are wholly comparable with the so-called electrochemical methods, or how and where they differ
2) Whether one method of analysis should be preferred to another, and if so, for what reasons.

It is for the sake of answers to these basic questions that I invited Professor Baudras of Toulouse, Dr. Munz of Boehringer, and Dr. Wuhrmann of Hoffmann-La Roche. Professor Baudras, thanks to his numerous studies in enzymology, particularly concerning LDH, will talk with his customary competence and authority on the theoretical aspect of enzymatic dosage. He will review some thermodynamic notions to demonstrate how enzymatic oxidation of the lactate can be realised stoichiometrically and employed to obtain an accurate measurement.

Dr. Wuhrmann and Dr. Munz will present the results of the work of their respective groups in the field of electrochemical and spectrophotometric dosage with the thoroughness and honesty inherent in true scientific research.

Chapter 9

Experience with an Enzymatic Monotest for the Determination of Lactate

E. Munz

Background

The knowledge of the concentration of lactate is of importance as a supplement to blood gas analysis and to acid-base status in various pathological situations, e.g., serious acidoses, states of shock, intensive care, perinatal medicine, and tolerance tests.

Interest in devising a rapid and simple test increased as appreciation of the diagnostic and prognostic value of lactate grew. Earlier test methods were seldom able to fulfil expectations, being relatively complicated and time-consuming.

Development of New Test

With a newly developed enzymatic test based on a publication of F. Noll (1), it was possible to establish a method that satisfies the demands of the routine laboratory.

The test is based on the following reaction:

$$\text{L-lactate} + NAD^+ \underset{}{\overset{LDH}{\rightleftharpoons}} \text{pyruvate} + NADH + H^+$$

$$\text{Pyruvate} + \text{L-glutamate} \overset{GPT}{\rightleftharpoons} \text{alanine} + \underline{\alpha}\text{-ketoglutarate}$$

The enzyme LDH catalyses the oxidation of lactate to pyruvate in the presence of NAD^+, with formation of NADH. The extinction increase of NADH, measured at 334, 340, or 365 nm, corresponds to the lactate content of the sample.

WORKING REAGENT
L-Lactate

$NAD^{\oplus}$	4.6 mmol/litre
Carbonate buffer	0.5 mol/litre
	ph 10.0
L-glutamate	63.0 mmol/litre
LDH	32 U/ml
GPT	2 U/ml

Fig. 1. Substrate and enzyme concentrations used in the assay

The problematical nature of this reaction is due to the fact that the equilibrium lies almost completely on the side of lactate and that the conversion with time proceeds very slowly. These difficulties could be eliminated by various means.

In the assay described in this paper, the substrate and enzyme concentrations shown in Fig. 1 were used. The improvements to the method can be attributed to the following points.

1) A large excess of NAD^+ is used
2) The reaction proceeds at an alkaline pH, at which the H^+ ions formed are removed from the reaction mixture
3) The pyruvate formed is converted to alanine in a coupled second reaction with GPT and glutamate, and is thus also removed from the reaction
4) By increasing the LDH and GPT concentrations in the assay mixture it is possible to shorten the reaction time to 10-15 min
5) Carbonate ions cause an additional activity increase of the GPT reaction.

The method is suitable for the determination of lactate in deproteinised blood and in plasma.

The strongly alkaline environment leads to a nonspecific extinction increase that is attributable to an instability of NAD^+ in this pH

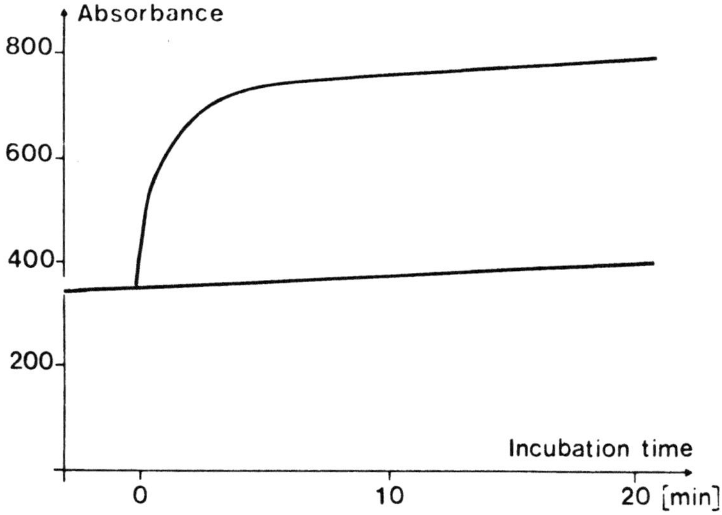

Fig. 2. Nonspecific extinction increase caused by the instability of NAD^+ in the alkaline environment used

range (Fig. 2). Since plasma is also evaluated as sample material, in general turbidity can also simulate an extinction increase. To eliminate these effects it became necessary to run a sample blank. The extinction difference of the method is 0.060 for a lactate concentration of 1 mmol/litre at a wavelength of 334 nm. The calculation is made via a constant factor. With plasma as sample material the reaction is completed after 10 min; when deproteinised blood is used the reaction time is increased by the perchloric acid deproteinator, which lowers the pH in the reaction mixture and slows the reaction correspondingly.

Early Experience with the Test

The first experiences and results with the method described here were obtained with a reagent kit built up on the basis of Monotest. The kit contains 19 individual bottles with the coenzymes, one bottle of ready-made buffer, and one bottle with the enzymes. The contents of a Monotest bottle are dissolved in 5 ml buffer. The stability of reagent solutions thus prepared is limited: 4 h at room temperature and at least 12 h at 4°C. This means that tests that have to be performed on night duty no longer require special preparation. The

Procedure - 334 nm/340 nm

Pipette into test tubes		
	Sample blank	Sample
Reagent solution	1000 µl	1000 µl
Plasma	10 µl	10 µl
Mixing		
Enzymes	–	20 µl
Water	20 µl	–
Mixing, incubation 10 min RT $\Delta E = E_S - E_{SB}$		

Calculation

	334 nm	340 nm
c [mmol/l]	16,7 x ΔE	16,3 x ΔE

Fig. 3. A pipetting scheme suitable for the assay

measurements can be carried out at wavelengths 334, 340, or 365 nm, both on macro and on semi-micro scales.

An example of a pipetting scheme is shown in Fig. 3. The extinction difference between sample and sample blank remains constant for a long period. It should be noted, however, that sample and sample blank should be measured in immediate succession.

Accuracy

The serial precision was determined in deproteinised controlsera, deproteinised full blood, and plasma with coefficients of variation from 1.8% to 4.6%. A good degree of precision was observed in all samples. No dependence on wavelengths was found. The day-to-day precision with coefficients of variation ranging from 1.8% to 4.0%

can be designated as excellent. The results were made available to us by Dr. Schwab of the Städtisches Krankenhaus, Karlsruhe.

The limits of measurement for the method were determined with deproteinisation and with plasma. A deviation from the linear part of the curve was found for the deproteinisation method at about 10 mmol/litre and for the plasma method at about 18 mmol/litre. The user can achieve a larger range of measurement, however, by increasing the incubation time. In my view, a useful practical compromise has been reached in the plasma method with a 10-min incubation time and a measurement limit of 18 mmol/litre.

Evaluation of Correctness of Results

Assessment of the correctness of a method frequently causes difficulties when no practicable reference method is available. Therefore investigations on correctness can often only be carried out by recovery tests.

In the first series of experiments the correctness was investigated by means of recovery experiments in aqueous lactate solutions. The exact amount of lactate used can be recovered at a concentration of 20 mmol/litre. With a lactate content of 30 mmol/litre the deviation from the reference value is -6%; at 40 mmol/litre it is about -10%.

Deproteinised blood plasma samples were supplemented with aqueous lactate solutions in various concentration steps. The recovery was determined at wavelengths 334 and 365 nm.

The most effective method is deproteinisation of full blood at the bedside. For quantitative precipitation of protein the perchloric acid must be cool. Since the free-flowing blood is supposed to run into the centrifuge tube it is hardly possible to achieve reproducible volume proportions with this procedure.

The collection of plasma has proved to be a considerably more practicable method. EDTA is used as an anticoagulant. Glycolysis is inhibited by the addition of potassium fluoride (KF), which inhibits the enzyme enolase. It only remains therefore to prepipette one drop of EDTA-KF solution per millilitre of blood into a centrifuge tube and to allow the blood to flow in. The EDTA-KF solution is available in a dropper bottle. A volume correction is not necessary in practice. The stability of the lactate concentration without centrifugation of the erythrocytes is about 2 h at room temperature. After centrifugation the lactate concentration remains stable at room temperature for about 3 days.

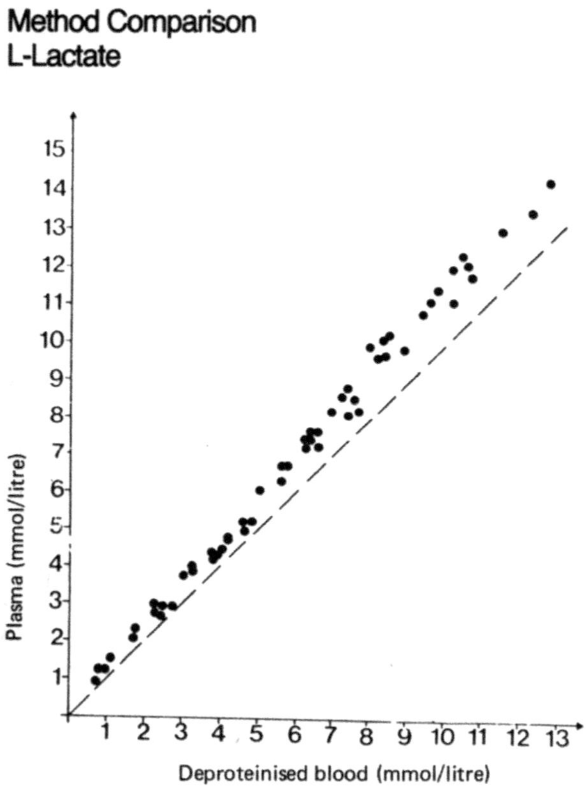

Fig. 4. Comparison of results obtained when assay is performed in plasma and in deproteinised blood from 53 patients. $\underline{y} = 1.13\underline{x} + 0.12$; $\underline{r} = 0.996$

Comparison of Results Obtained in Plasma and in Deproteinised Blood

The methods with deproteinised blood and with plasma were compared on samples from 53 patients (Fig. 4). As can be seen from the data, the plasma method yields values approximately 13% higher. The cause of this effect was determined. During the deproteinisation with perchloric acid the erythrocytes are destroyed and their contents reach the sample. Since in erythrocytes the lactate content is only 15%-20% of the plasma concentration a dilution effect occurs. The true lactate content is therefore only measured in plasma. The use of centrifuge tubes pretreated with KF and EDTA is also possible. These tubes are available from various manufactures in a ready-to-use form.

In all samples the weight of lactate was recovered quantitatively; the recovery mean was 99%.

The recovery was also tested in various control sera available on the market. The data show that the reference values given by the manufacturers for lactate were found very accurately, the range of recovery varying from 95% to 104% of the expected values. It is thus assured that an internal laboratory quality control can be carried out with the lactate method described here.

Specificity

In further experiments the specificity of the method was investigated. Pathological concentrations of metabolic products such as lipoproteins, bilirubin, haemoglobin, pyruvate, and 3-hydroxybutyrate did not influence the measured results. The 2s range of serial precision was taken for the deviation limits.

A total of 35 drugs from a wide range of therapeutic areas was added to blood samples. None of these additives disturbed the method. The above experiments show clearly that reliable results can be obtained with the method introduced here for lactate measurement.

While the actual lactate determination presents no problems in practical terms, the isolation and preparation of plasma must be carried out with special care. The collection of the sample material and its stability should be discussed in more detail.

After blood collection, the lactate content increases continuously due to glycolysis, and in samples from the normal range reaches a value about 20% above the initial value after 30 min. It is therefore essential to interrupt the glycolytic process.

In the medical clinic at Tübingen University the determination of lactate in capillary blood is being investigated. A further simplification of the blood cellection can be observed here, since now only 20 μl blood is required for one determination.

Summary

Lactate determination has a high predictive value, especially in the diagnosis of acidic metabolic disorders. Up to now the application of the test has been hampered by its inconvenience. We present a UV test for quick, convenient, and reliable L-lactate analysis, based on the LDH-catalysed oxidation of L-lactate to pyruvate. The determination can be performed in plasma or in deproteinised blood. This method requires a consecutive reaction step, which in this case is the removal of pyruvate by an ALT-catalysed glutamate-pyruvate transamination. The reaction is complete within 10 min at room temperature. The precision in series varies from 1.8% to 4.6%, depending on the lactate concentration; the linearity range is up to 18 mmol/litre in plasma and 10 mmol/litre in deproteinised blood; the mean recovery after the addition of L-lactate to sera is 99%. No interference was observed to result from drugs or metabolites. Due to the instability of NAD^+ in alkaline solutions, the stability of the reconstituted assay solutions is limited to 4 h at room temperature and 12 h at +4°C. The test is conceived as a Monotest.

References

BERGMEYER, H.U.: Methoden der enzymatischen Analyse. 3rd ed., pp. 1510-1514, 1521-1525 (1974)
NOLL, F.: Biochem. Z. 346, 41 (1966)
HOHORST, H.J.: Biochem. Z. 328, 509 (1959)

Chapter 10

Suitability of Yeast Lactate Dehydrogenases (Cytochrome b2) for Lactate Determination

A. Baudras

L (+) Lactate Oxidation Catalysed by Muscle or Yeast LDH: Thermodynamics

Insofar as the oxidation or reduction of the lactate-pyruvate system is to be operative, it is necessarily coupled to the reduction or oxidation of an appropriate acceptor redox system, whatever the enzyme used to catalyse the reaction.

The extent to which lactate may be oxidised, or alternatively, the extent to which pyruvate may be reduced, depends directly on the relative values of the half-reduction potentials of the two systems present:

Lactate/Pyruvate ‖ $Acceptor_{ox.}/Acceptor_{red.}$.

Scheme 1 shows the E_m values and the resulting $K_{eq.}$ (Lact. → Pyr.) at pH 7 and 25°C. From these data, the following conclusions are readily apparent: (1) When muscle LDH is to be used as a catalyst, the obligate acceptor is NAD^+, the coenzyme of LDH_M. Low E_m (negative) of the coenzyme acceptor system makes it act spontaneously as an electron

SCHEME 1

$$\text{L(+)lactate} + NAD^+ \underset{}{\overset{LDH_M}{\rightleftharpoons}} \text{Pyruvate} + NADH + H^+ \qquad (1)$$

$$\text{L(+)lactate} + 2\,[Fe(CN)_6]^{3-} \overset{LDH_Y}{\rightleftharpoons} \text{Pyruvate} + 2\,[Fe(CN)_6]^{4-} + 2H^+ \qquad (2)$$

	NAD^+/NADH		LACT./PYR.		FERRI./FERRO.
$E_{m,7}$ (Volt)	-0.32		-0.19		+ 0.36
$K_{eq.}$ (LACT.→PYR.)		≅ 5.10^{-5} (1)		≅ 3.10^{18} (2)	

$$[\Delta G' \text{ (pH 7)} = -nF\cdot\Delta E_{m.7} = -RT \ln K_{eq.}]$$

donor, and the reaction pyruvate → lactate is spontaneous. At pH 7, LDH_M is thus an excellent pyruvate reductase. However, it can also be made a good lactate oxidase by displacing the reaction equilibrium to the right by "product trapping" i.e., trapping the protons by making use of a high pH buffer (pH > 9) and trapping pyruvate produced by forming a very stable compound with hydrazine. Another general way of achieving this last result would be to have pyruvate transformed in any kind of spontaneous reaction (enzymatic or not). (2) By contrast, yeast LDH is unreactive towards pyridinium coenzymes, but transfers electrons very efficiently to ferricyanide (or various dyes with positive E_m values). The acceptor here having a quite high and positive $E_{m,7}$ value, $K_{eq.}$ (Lact. → Pyr.) is very high and lactate oxidation is almost irreversible near neutrality.

Insofar as the condition of a high value of $K_{eq.}$ (Lact. → Pyr.) is fulfilled, a stoichiometric consumption of lactate can be achieved whenever the oxidised form of the acceptor system is present in excess of the amount of electrons to be withdrawn from lactate contained in the sample. The enzyme only accelerates the process of going to equilibrium, which is characterised by a near-zero equilibrium concentration of lactate. Not only is the excess of oxidised acceptor necessary for completion of lactate oxidation, but also it is the very basis of the simplest quantitative determination procedures, which rely on the measurement of the change between the initial and final concentration levels of oxidised (or reduced) acceptor. To avoid overstepping the "electron capacity" of the oxidised acceptor and its complete consumption, appropriate dilution of the sample is necessary. The amount of oxidised acceptor consumed at equilibrium (hence the amount of lactate present in the sample) is commonly computed from spectrophotometric measurements at selected wavelengths (340 nm for formation of NADH; 420 nm for reduction of ferricyanide to almost colourless ferrocyanide). These measurements are very sensitive, very accurate, and practically instantaneous, not counting the time needed for pretreatment of the samples or the time needed for the reaction to be completed in the test cuvette.

Yeast Lactate Dehydrogenases (Cytochrome b_2)

When grown aerobically, baker's yeast (*Saccharomyces cerevisiae*) is known to contain the flavin-heme protein L(+) lactate:ferricytochrome c oxidoreductase (EC 1.1.2.3), which is also called cytochrome b_2. This type of enzyme is considered to be inducible and associated with mitochondria, and to catalyse the oxidation of L(+) lactate and some other α-hydroxy acids with a coupled electron transfer to cytochrome c in vivo. This protein is not very tightly associated with the mitochondrial membrane, since it is rather easily solubilised compared with the classic b-type cytochrome of the respiratory chain.

Different forms of the enzyme are currently available in the pure state, extracted either from baker's yeast or from a strictly aerobic organism, the yeast Hansenula anomala. In any case, the pure active protein was shown to be a tetramere (molecular weight 230 000) with four flavins (FMN) and four protoheme prosthetic groups non covalently associated with the polypeptide chains. A few years ago, it became apparent that the crystalline classic form of the Saccharomyces enzyme had in fact suffered a limited proteolysis with chain cleavage during the purification (1, 2). Modified and active preparations of this kind are now currently termed "cleaved" forms of cytochrome b_2, as opposed to the corresponding "noncleaved" or "physiological" forms. Selected properties of the various pure enzyme forms are presented in Table 1. It is worth noting that Hansenula enzyme combines high molecular activity, excellent stability, and unmatched yield during purification (recoveries as high as 1 g pure enzyme per kg dry weight of yeast have been recently obtained with excellent yeast cultures (3). An up-to-date evaluation of the purification procedures together with selected references on structure and properties of cytochrome b_2 can be found in Methods in Enzymology (4).

Though intrinsically stable, preparations of cytochrome b_2 need to be handled with due caution to avoid their contact with inactivators of various kinds (considered briefly below). Detergents and general protein-denaturing agents (urea, guanidium chloride, strong oxidants, etc.) must be avoided, as must specific mercurials or other sulphydryl groups, reagents, and metal ions or complexes, particularly Cu^{2+} and Ag^{+}. A very important problem concerns traces of proteases, whether already present as contaminants in certain enzyme preparations or introduced by the use of improperly cleaned glassware and during experimentation, for example through bacterial contaminations. This very serious problem originates in the fact that there are regions of the polypeptide chain of cytochrome b_2 that are readily accessible and sensitive to proteases (5, 6), as exemplified by the occurrence

Table 1. Some characteristic properties of yeast L(+)lactate dehydrogenases (cytochrome b_2)

	K_m (lact.)[a] (m*M*)	Molec. activity[a] (s^{-1})	K'_S[a, b] (M)	Stability	Purific. yield (mg/kg dry weight)
Saccharomyces cerevisiae					
Crystalline (cleaved)	1.3	235	0.3	Good	35-50
Physiological (noncleaved)	0.4	400	0.13	Mean	35-50
Hansenula anomala					
Crystalline (noncleaved)	1.3	1300	0.04	Excellent	200-1000

[a] From A. Baudras, 1973
[b] K'_S is the constant for inhibition by excess substrate

of the so-called cleaved from of _Saccharomyces_ enzyme. Other problems concerning stability may be more specific and related to peculiar structural features or reactivity of the enzyme forms considered. Thus, _Hansenula_ enzyme is inactivated quite rapidly when exposed to low ionic strength buffers ($I < 0.1$ _M_) at room temperature due to a dissociation (reversible) of the active tetramere into inactive monomers (7, 8). This same enzyme is susceptible to photo-inactivation, which is currently being investigated (B. Blazy and A. Baudras, unpublished observations). In particular, dissociable complexes formed between the enzyme and a class of effectors including pyruvate and oxalate may be converted into inactive covalent compounds by flavin adduct formation on exposure to light, in a way very similar to that already described for the flavin-enzyme lactate oxidase of the bacteria _Mycobacterium smegmatis_ (9).

Spectrophotometric Determination of L(+) Lactate by Means of Yeast LDH

The use of yeast LDH for L(+) lactate determination was suggested many years ago (10). However, renewed interest in such a method (11) emerged only recently, with the availability of very active and stable enzyme preparations. Spectrophotometric measurement of ferricyanide reduced on irreversible oxidation of L(+)lactate mediated by cytochrome b_2 is the basis for a very simple and accurate determination.

With normal human blood samples, no discrepancy was observed between values obtained with this procedure and those obtained with the common lactate-UV test (muscle LDH and NAD^+) (11). However, in both cases blood samples need to be deproteinised prior to lactate measurement, this step being by far the most time-consuming operation with this kind of procedure. As shown in Fig. 1, the reaction is complete within a few minutes (at 25°C and pH 7.2), at the expense of some 5-10 μg pure enzyme per test cuvette. Note that because the reaction is nearly irreversible, doubling the amount of enzyme reduces this response time roughly by a factor of 2, so that a very fast response can be achieved at the expense of more enzyme. Rapid completion of the reaction together with visible photometric measurement (at 420 nm) and use of inexpensive ferricyanide as an electron acceptor are distinct advantages of the method. Interference with substrates related to lactate may arise with certain biological samples, since the spec-

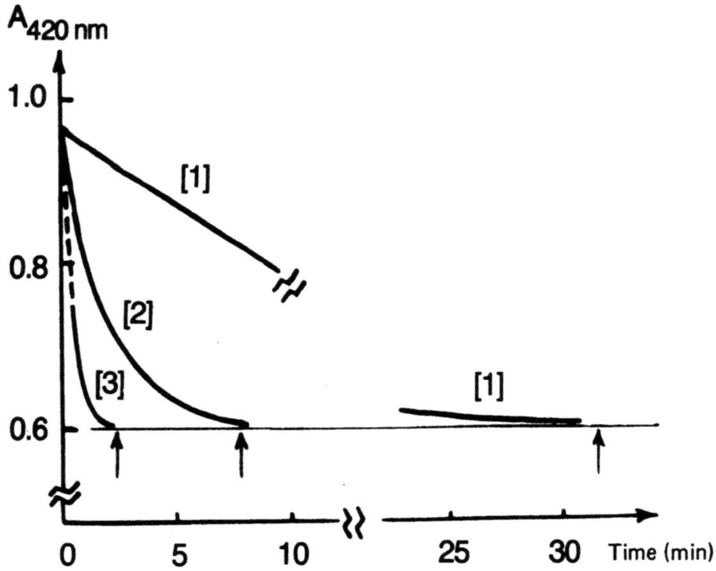

Fig. 1. Progress curves for L(+)lactate oxidation by ferricyanide in the presence of yeast LDH. Absorbance of ferricyanide at 420 nm was recorded as a function of time. Reactions proceed at 25°C and pH 7.2 with the following amounts of LDH_y per cuvette: (1) 0.55 U or 1 μg; (2) 2.2 U or 4 μg; (3) 7.7 U or 14 μg (pure enzyme is 550 U/mg). Deproteinised and diluted blood samples were prepared as described by H. Durliat et al., Clin. Chem. 22: 1802, 1976

ificity of yeast LDH for L(+)lactate is not absolute. Some aspects concerning selectivity will be dealt with below, being relevant to both spectrophotometric and electrochemical-enzymatic determinations.

L(+)Lactate Determination by Means of an Enzyme Electrode

Lactate measurements with enzyme electrodes were first described in principle by Williams et al. (12), and later improved by Roche's group in Basel (13) to the present state of a semi-automatic "lactate analyzer" (14). Basic investigations were also conducted independently at the University of Toulouse, with a view to optimising critical variables of the electrochemical-enzymatic sensor (11, 15).

The active part of the device is the enzyme electrode, shown schematically in Fig. 2. It comprises a very small enzyme chamber limited at one end by a platinum disc and in contact through a cellophane membrane with the sample solution. The enzyme, yeast LDH, cannot escape from the chamber, whereas small molecules diffuse quite easily

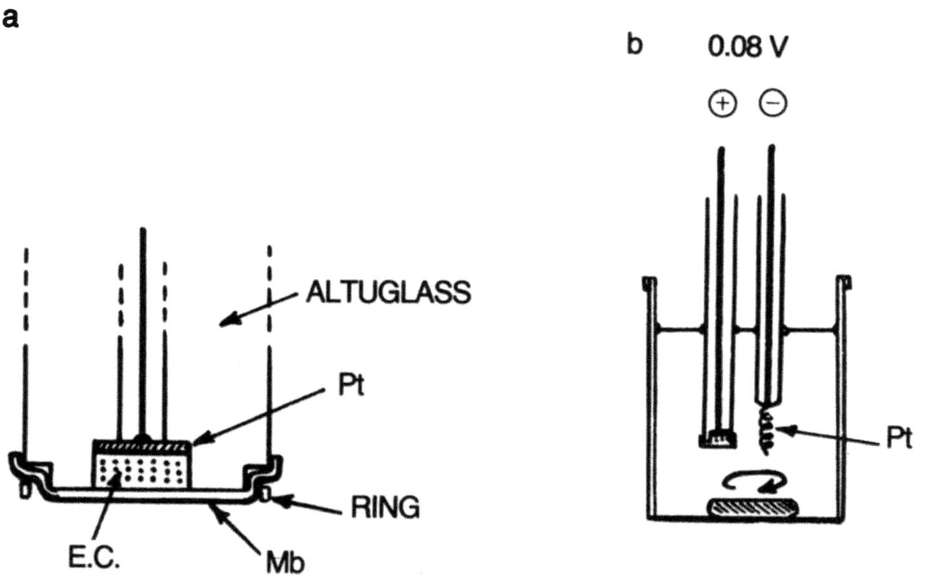

Fig. 2. a and b The lactate sensor and set-up for electrochemical-enzymatic determination of L(+)lactate. a The active part of the enzyme electrode: E.C. is the enzyme chamber, and Mb the semipermeable membrane. b A scheme of the titration vessel and electrode set-up; the DC-regulated power supply delivering a constant potential of 0.08 V, and the electronic equipment for measurement of current *i* are not shown. Typical values observed in such measurements with 1.0 m*M* L(+) lactate (25°C; pH 7.2) are: i_{max} = 1.5 μA and θ = 24 s for the following values of the enzyme electrode characteristics: 0.4 μ*M* $\leqslant$ (E_o) < 130 μ*M*; chamber depth *d*, 40 μm; membrane thickness *e*, 50 μm

through the semipermeable membrane. The sample solution containing lactate being supplemented with ferricyanide (1-2 m*M*), the reacting species (L(+)lactate and ferricyanide) enter the chamber by diffusion and both the following reactions take place:

$$1(+)\text{lactate} + 2\,[Fe(CN)_6]^{3-} \xrightleftharpoons{LDH_Y} \text{pyruvate} + 2\,[Fe(CN)_6]^{4-} + 2\,H^+$$

$$[Fe(CN)_6]^{4-} \xrightarrow[Pt]{} [Fe(CN)_6]^{3-} + e^-$$

The second reaction, which is purely electrochemical and limited to the platinum surface, is responsible for a current *i*, to be measured when the enzyme electrode is properly coupled to a second electrode - the cathode - and a constant and selected voltage is applied between them. As soon as the electrodes are dipped in the titration vessel, the time profile of current *i* is observed experimentally to increase until a maximum value (i_{max}), which may extend as a plateau when stirred and/or large volumes of sample solution are used (12, 13, 15). The time elapsing up to this point (or plateau) (θ) is the response time of the set-up. At this precise point (or along this plateau), a steady state is established between diffusion and reaction fluxes. Both $\underline{i}_{max}$ and θ are thus expected to depend on enzyme concentration, factors that control diffusion (chamber depth and membrane thickness), and the external concentration of L(+)-lactate (ferricyanide may be kept constant and in any case it is regenerated during operation).

Different choices of electrode set-up and working potential difference have been proposed since the pioneering work of Williams' group (12), but it was H. Durliat and M. Comtat who optimised the critical variables of the enzyme electrode (15) and showed that a simple platinum wire may be used as a cathode provided that a working potential of 0.08 V is used (delivered by a DC-regulated power supply) (11). An optimised set-up of this kind permits response times shorter than 1 min, together with an extended range of linear response of $\underline{i}_{max}$ versus lactate concentration (0.1 m$\underline{M}$ to about 10 m$\underline{M}$) (11, 15). Critical variables are the enzyme concentration E., the depth $\underline{d}$ of the enzyme chamber, and the membrane permeability or its thickness $\underline{e}$ (for a given type of membrane). These different factors were shown (15) to modulate the response of the electrode as follows: (1) increased values for response time θ together with decreased values for $\underline{i}_{max}$ are observed when $\underline{e}$ and/or $\underline{d}$ are increased (reduced diffusion fluxes); (2) both θ and $\underline{i}_{max}$ are strictly independent of (E_o) over a very large range of enzyme concentration ((E.) $\geqslant$ 0.4 $\mu\underline{M}$ or 24 μg/ml). In these conditions, lactate is oxidised so rapidly that its steady-state concentration within the chamber is near zero. The first of these two points is of practical interest, as it leads to the use of chambers with a very small volume (a few microlitres) to ensure very small values of $\underline{d}$, with the benefit of both fast response and high sensitivity (high $\underline{i}_{max}$ value). The second point has very important implications. On the one hand, progressive inactivation of any origin has no effect on the electrode response so long as the true concentration in active enzyme remains higher than about 0.4 $\mu\underline{M}$, and this ensures a long life of the electrode (usually 3-4 weeks with no need to renew the enzyme solution in the chamber). Moreover, calibrations with standard solutions of authentic L(+)-lactate need not be frequent (in the absence of electronic drift). On the other hand, it is predictable that working with a high (E.) will render reversible inhibitors ineffective, as far as the electrode response is concerned, even if they are present in quite high concentrations (see below).

Selectivity: Effects of Reversible Inhibitors and Alternative Substrates

There are two reasons why problems may theoretically be encountered when LDH is used for L(+)lactate determination in biological fluids. One is connected with the presence of lactate analogues, some of which may be reversible inhibitors of LDH_y. The other is to do with the fact that LDH_y (extracted either from *Saccharomyces* or from *Hansenula*) is known to accept a few alternative substrates other than L(+)lactate, all of which are L(+) α-hydroxy acids, whereas D-isomers are generally inhibitors of the enzyme LDH_y.

Problems of the first kind are not serious, because basically, reversible inhibitors prevent a certain fraction of the enzyme molecules from working thus increasing reaction times, but in any case this effect is overcome by using more enzyme. As already discussed, provided it works with a high (E_o) within the enzyme chamber, the enzyme electrode can in principle accomodate very large concentrations of inhibitors without any change in its response ($\underline{i}_{max}$ and θ). This is exemplified by observations reported in Fig. 3, which show titrations of mixtures of L(+) and D(-)lactate. Traces of L(+)-lactate as low as 1%-2% in commercial preparations of D(-)lactate are quite readily determined. The data shown correspond to a D(-):L(+)-ratio of about 50 and to values of D(-)lactate concentration up to about 10 K_i (K_i is the dissociation constant of the enzyme-inhibitor complex). The curvature observed, together with the slight difference in slope at the origin, suggests that the concentration of enzyme in the electrode may not have been sufficient here - due to the high D(-):L(+)ratio - to ensure $\underline{i}_{max}$ was still independent of $(E_o)_{active}$. If the measurements had been performed with a higher enzyme concentration in the electrode (say 100 $\mu\underline{M}$ instead of 10 $\mu\underline{M}$), a more extended range of linear response would have been observed and the two straight lines would presumably have become parallel. The actual concentration of L(+)lactate in the D(-) sample may be estimated from the value of total concentration (weight basis), which gives an indicator signal identical to that observed with a known concentration of authentic L(+)lactate. With strictly pure D(-)lactate no

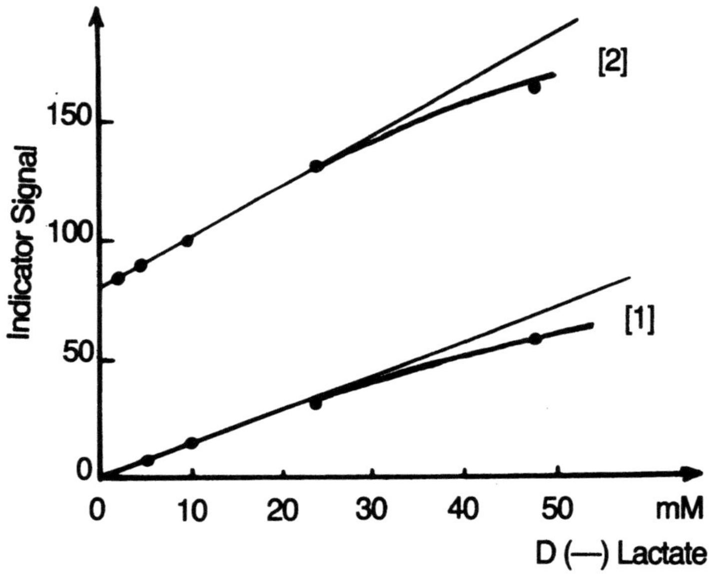

Fig. 3. Determination of L(+)lactate with the lactate sensor in the presence of high concentration of D(-)lactate. The indicator signal (proportional to $\underline{i}_{max}$) is given as a function of total D(-)lactate concentration (Sigma L-0625) either alone (1), or in the presence of a constant 1.0-mM concentration of authentic L(+)lactate (2). Measurements were performed at 20°C, pH 7.2; the enzyme concentration in the electrode (E_0) was 10 µM. The amount of L(+)lactate present in the D(-) sample is estimated to be 2%-3%

signal would be observed, since this compound is not a substrate of LDH_y but an inhibitor (16).

Problems of the second type may generally be more serious. Not only is it predictable that substrates other than L(+)lactate will be measured by the enzyme electrode, but furthermore, it is our "central hypothesis" that provided (E_0) is very large and the diffusion coefficients of the supposed substrates have comparable values, any of them should give the same indicator response for a given concentration. As shown in Table 2, the alternative substrates of LDH_y (from Hansenula) known at present are "poor substrate", being oxidised with a much lower molecular activity than L(+)lactate itself. The possibility of titrating substrates other than L(+)lactate with the so-called lactate electrode is illustrated in Fig. 4 in the case of phenyllactate, a very poor substrate. With a very high concentration of enzyme in the electrode (here 115 µM), a nicely linear correlation is observed between the indicator signal read and the total phenyl-

Table 2. Substrates[a] interfering with L(+)lactate determination

	K_m (approx.)	Molec. activity (s^{-1}; approx.)	Sensor response
2-OH butyrate	2.8 mM	72	Yes
Phenyllactate	0.6 mM	26	Yes
3-OH Butyrate	No binding		Not detected
2-OH Isovalerate	Inhibitor K_i (approx.) 3.5 mM		Not detected

[a] All substrates tested here were commercial DL samples. (E_o) in the sensor was 115 μM

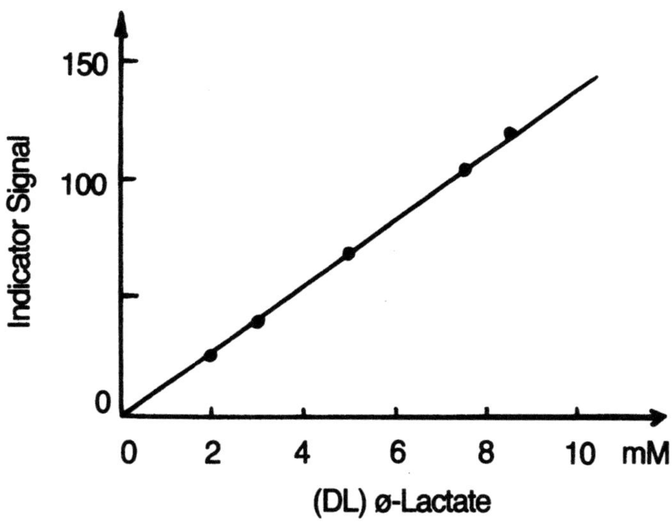

Fig. 4. Determination of phenyllactate with the lactate sensor. The indicator signal (proportional to i_{max}) is given as a function of total phenyllactate concentration (DL-phenyllactate Sigma P-7251). Measurements were performed at 20°C, pH 7.2; the enzyme concentration in the electrode (E_o) was 115 μM. With the calibration of the indicator used, a 1.0-mM solution in authentic L(+)lactate would read 40. The D:L ratio for this sample may be estimated to be ≅ 1.3/0.7 (see text)

lactate concentration, and the response time (θ) is identical with that observed when L(+)lactate is determined with the same electrode. Within our central hypothesis and with the procedure outlined above for estimation of traces of L(+)lactate in D(-) samples (Fig. 3), a value may be computed for the D:L ratio. Independent direct measurements of L(+) or D(-) isomer concentration in the sample have not yet been performed. Extending such comparative measurements to

D-L mixtures in the case of other substrates with known diffusion coefficients would allow the validity of the hypothesis to be checked.

A somewhat optimistic note is appropriate at this point, which is justified on theoretical grounds when both simple enzyme kinetics and the properties of the lactate sensor are considered. The idea is that it is possible to exploit the very low molecular activity of the enzyme for poor substrates to obtain selectivity when L(+)lactate is to be determined with the enzyme electrode in the presence of other substrates. It should be possible to achieve this by use of low (E_o) in the lactate electrode, down to the range in which θ and $\underline{i}_{max}$ become dependent on enzyme concentration: the lower the molecular activity for a given substrate, the wider difference is observed when the value for $\underline{i}_{max}$ is compared with that obtained with the same concentration of L(+)lactate. Table 3 illustrates this property; the signal observed with phenyllactate was much more strongly depressed than that observed with L(+)lactate at low (E_o) values. However, because of this, θ is increased and, most importantly, the buffering effect of large values of (E_o) is lost, so that the stability of the enzyme preparation used becomes very critical, as does the need for frequent calibrations.

Interference with substances that could be oxidised directly on the platinum part of the enzyme electrode or could reduce ferricyanide has not been alluded to in this presentation. Background currents with such an origin are easily recognised, since they are still present in the absence of enzyme in the electrode (or with inactivated enzyme).

As far as selectivity is concerned, no problems have been encountered up to now in the determination of L(+)lactate in whole or diluted human blood (11, 14, 15, 17). With any other kind of sample (wines, fermentation, or culture fluids, etc.), it is most desirable first to try to find out whether other substrates of the enzyme than L(+) lactate might possibly be present in the samples to be investigated. Finally, it should be stressed that problems of selectivity are relevant to both spectrophotometric and amperometric (enzyme electrode) measurements.

Table 3. Increased selectivity with low (E_0) in the lactate sensor[a]

(E_0) µM	Response time	L(+)lactate 1.0 mM (1)	DL-Ø lactate 1.0 mM (2)	L(+)lactate 1.0 mM + DL-Ø lactate 1.0 mM	(1) + (2)	(1)/(2)
10	10 s	42.5	14.7	58.5	57.2	2.9
1.0	90 s	43.5	12.0	56.0	55.5	3.6
0.26	4 min	45.0	12.0	58.5	57.0	3.75
0.10	6 min	22.5	2.5	23.5	25.0	9.0

[a] Figures show indicator response (proportional to $\underline{i}_{max}$)

Acknowledgements

The author is indebted to Dr. Hélène Durliat who contributed the experiments (unpublished) shown in Figs. 3 and 4 and Table 3.

Summary

Simple thermodynamic principles are recalled for the purpose of discussing how an enzyme-mediated oxidation of lactate can be made stoichiometric and be used for its precise determination.

Lactate dehydrogenases from aerobic yeasts are briefly presented with reference to their structure, prosthetic groups, major catalytic properties, and availability.

Two suitable procedures for L(+)lactate determination are presented and discussed. One is a spectrophotometric method and the other is based on the use of the so-called lactate electrochemical-enzymatic sensor (lactate electrode); both use yeast lactate dehydrogenase as a catalyst.

Factors that can affect enzyme structure and/or interfere with the catalytic oxidation of L(+)lactate are considered, with particular attention to the problem of selectivity in L(+)lactate determinations.

References

1. LABEYRIE, F., BAUDRAS, A.: Differences in quaternary structure and constitutive chains between two homologous forms of cytochrome b_2 (L-lactate:cytochrome c oxidoreductase). Eur. J. Biochem. 25, 33 (1972)
2. JACQ, C., LEDERER, F.: Sur les deux formes moléculaires du cytochrome b_2 de Saccharomyces cerevisiae. Eur. J. Biochem. 25, 41 (1972)
3. BLAZY, B., BARDET, M., BAUDRAS, A.: A standard high yield purification of L(+)-lactate:cytochrome c oxidoreductase (cytochrome b_2) from the yeast Hansenula anomala. Anal. Biochem. 88, 624 (1978)
4. LABEYRIE, F., BAUDRAS, A., LEDERER, F.: Flavocytochrome b_2 or L-lactate:cytochrome c reductase from yeast. Methods Enzymol. 53, 238 (1978)
5. NASLIN, L., SPYRIDAKIS, A., LABEYRIE, F.: A study of several bonds hypersensitive to proteases in a complex flavohemoenzyme, yeast cytochrome b_2. Eur. J. Biochem. 34, 268 (1973)
6. POMPON, D., LEDERER, F.: Controlled proteolysis of flavocytochrome b_2. Eur. J. Biochem. 68, 415 (1976)
7. BAUDRAS, A.: Pure L-lactate:cytochrome c oxidoreductase of the yeast Hansenula anomala offers new prospect for the study of intramolecular multielectron transfer. Biochimie, 53, 929 (1971)
8. BAUDRAS, A.: Corrélations entre cinétique d'inactivation/réactivation et changements d'états de structure quaternaires du cytochrome b_2 de la levure Hansenula anomale, régulation par des effecteurs. In "Aspects dynamiques des changements de conformation dans les macromolécules biologiques" Edit. Sadron, C.; D. Reidel (Dordrecht), p. 181 (1973)
9. GHISLA, S., MASSEY, V.: Mechanism of inactivation of the flavoenzyme lactate oxidase by oxalate. J. Biol. Chem. 250, 577 (1975)
10. WIELAND, O.: An optical-enzymic determination of L(+)lactic acid with a diphosphopyridine nucleotide (DPN)-independent lactic acid dehydrogenase from yeast. Biochem. Z. 329, 568 (1958)
11. DURLIAT, H., COMTAT, M., BAUDRAS, A.: Spectrophotometric and electrochemical determinations of L(+)lactate in blood by use of lactate dehydrogenase from yeast. Clin. Chem. 22, 1802 (1976)
12. WILLIAMS, D.L., DOIG, A.R., KOROSI, A.: Electrochemical-enzymatic analysis of blood glucose and lactate. Anal. Chem. 42, 118 (1970)
13. RACINE, P., Mindt, W.: On the role of substrate diffusion in enzyme electrodes. Experientia [Suppl.] 18, 525 (1971)
14. RACINE, P., ENGELHARDT, R., HIGELIN, J.C., MINDT, W.: An instrument for the rapid determination of L-lactate in biological fluids. Med. Instrum. 9, 11 (1975)
15. DURLIAT, H., COMTAT, M., MAHENC, J., BAUDRAS, A.: Recherche des conditions optimales de fonctionnement d'une électrode à enzyme spécifique du lactate. Application au dosage dans le sang. Anal. Chim. Acta 85, 81 (1976)
16. BAUDRAS, A., SPYRIDAKIS, A.: Etude de la L(+)lactate:cytochrome c oxydoréductase (cytochrome b_2) de la levure Hansenula anomala. Biochimie 53, 943 (1971)
17. GUARNACCIA, R., RACINE, P.: Investigation on the specificity of an amperometric enzymatic sensor for L-lactate. Biomed. Techn. 21, 189 (1976)

Chapter 11

Amperometric – Enzymatic Analysis of Lactic Acid

H.R. Wuhrmann

Since the concentration of lactate in blood may vary rapidly, the usefulness of a lactate assay frequently depends on the speed of the analytical method. The enzymatic-amperometric technique described here offers the possibility of routinely determining the lactate level within a few minutes after sample collection.

The feasibility of a lactate sensor of the type described here was first demonstrated by Williams and co-workeers in 1969 (1). Their sensor had a small linear range form 0 to 2 mM/litre and was found to drift excessively. Racine et al. (2)-(4) later studied the factors determining linearity and drift behaviour of such enzyme electrodes with the purpose of improving this method.

The characteristic behaviour of a particular enzyme electrode depends on several parameters, such as diffusion processes, kinetics of the enzyme reaction, and the type of electrochemical reaction.

Three types of electrodes can be distinguished, according to what is being measured:

1) A product of the enzyme reaction
2) The consumption of a coenzyme or acceptor determined from the decrease of its initial concentration
3) The amount of the converted, reduced, or oxidised form of a coenzyme.

It will be shown that under optimal operation conditions the third type of enzymatic sensor is suitable for the determination of lactate in biological samples.

The enzymatic sensor consists of a disc-shaped, slightly recessed platinum anode coated with a thin layer of dissolved enzyme. The enzyme is separated from the test solution by a semipermeable mem-

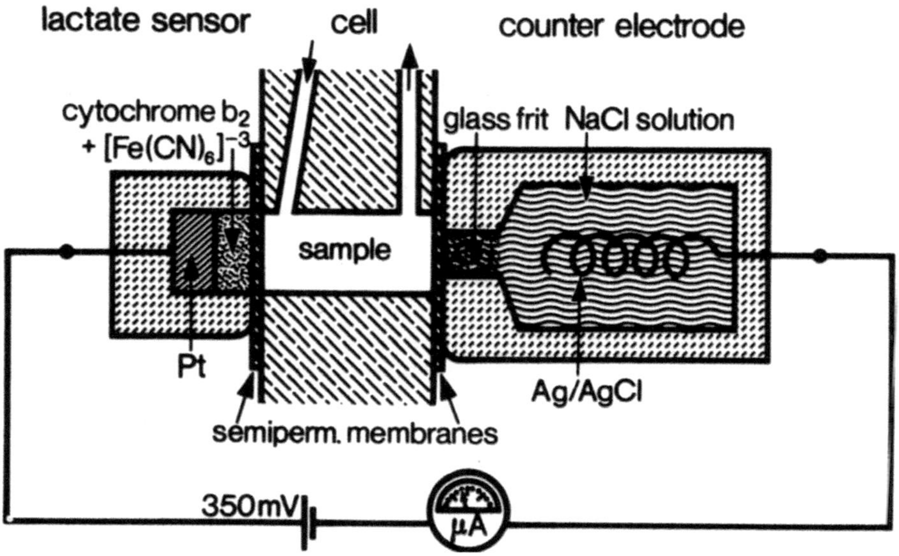

$$\text{lactate} + 2[Fe(CN)_6]^{-3} \xrightarrow{\text{cyt. } b_2} \text{pyruvate} + 2[Fe(CN)_6]^{-4} + 2H^+$$

$$2[Fe(CN)_6]^{-4} \xrightarrow{Pt} 2[Fe(CN)_6]^{-3} + 2e^-$$

Fig. 1. The principle of the measurement is based on the highly specific oxidation of lactate to pyruvate in the presence of the enzyme cytochrome b_2 and hexacyanoferrate-(III). The hexacyanoferrate-(II) produced in this reaction is reoxidised at the platinum anode, giving two electrons for each converted lactate molecule (cf. Fig. 1). Thus, the current measured is related to the lactate concentration in the solution

brane, which is permeable to the low-molecular-weight compounds participating in the reaction, but impermeable to the enzyme.

For easy use in a clinical environment, the sensor should have the following properties:

The steady-state current must be proportional to the lactate concentration of the samples, which can be as high as 20 mmol/litre in undiluted samples.

The calibration line must exhibit only minor drift with time, despite unavoidable enzyme denaturation.

Finally, the response time of the sensor must be less than 2 min.

To see how these criteria can be satisfied, a short look into a theoretical model of the enzyme electrodes in necessary. To simplify the model, the following assumptions are made:

1) The concentration of the coenzyme in the enzyme layer is sufficiently high for the enzyme reaction to proceed under saturated conditions with regard to the coenzyme
2) The enzyme layer is very thin, so that the concentration of the substrate within this layer can be assumed to be uniform
3) The diffusion time of the reduced ferrocyanide to the surface of the platinum electrode is short compared with the diffusion time of lactate through the membrane
4) There is also negligible loss of ferrocyanide, due to diffusion through the membrane into the external solution.

Between measurements the cell is filled with a rinsing buffer containing hexycyanoferrate-(III), and the enzyme layer is therefore equilibrated with hexacyanoferrate-(III). To perform a measurement the cell is emptied and then the sample is injected. The first step is then the diffusion of the substrate lactate through the semipermeable membrane into the enzyme layer. If the current density is low, one can ignore the effect of a stagnant layer at the external side of the membrane, and the molecular flow of lactate through the semipermeable membrane is in accordance with the equation:

$$\left(\frac{dn}{dt}\right)_{diff} = A \cdot P_s \; (c_s^{ex} - c_s^{in}) \tag{1}$$

P_s denotes the membrane permeability to lactate and A the area of the exposed membrane.

With the assumption of coenzyme saturation the second step can be considered as an enzyme reaction according to Michaelis-Menten kinetics:

$$\left(\frac{dn}{dt}\right)_{ER} = -\frac{k_2 \; N_E}{1+\frac{K_M}{c_s^{in}}} = -\frac{i}{2F} \tag{2}$$

In this case the reduced enzyme is immediately oxidised by reduction of the ferricyanide to ferrocyanide. This determines the rate of substrate consumption in the enzyme layer, where N_E is the amount of enzyme and K_M the Michaelis constant (2, 3).

In the third step the ferrocyanide is quantitatively oxidised at the anode. This third reaction has to be fast compared with reactions 1 and 2.

In the steady-state condition the amount of lactate that diffuses into the enzyme layer is equal to the amount of lactate converted by the enzyme reaction; and since there are two electrons liberated with each decomposed lactate molecule this rate is equal to the current divided by twice the Faraday constant.

The model shows that for a linear response of the sensor the lactate concentration in the enzyme layer has to be much smaller than the Michaelis constant. In this case, the enzyme reaction proceeds according to first-order kinetics, and the current i measured at the platinum electrode is then linear with respect to the external lactate concentration, but still a function of the amount of enzyme in the electrode layer, N_E:

$$i = 2\,FA \frac{P_s \cdot c_s^{ex}}{1 + \dfrac{AP_s}{k_1\,N_E}} \qquad (3)$$

This is undesirable in cases in which the enzyme loses its activity by denaturation with time, and therefore makes frequent recalibration of the electrode necessary. One may see that the current becomes independent of N_E if the permeability of the membrane is made sufficiently low, so that the quantity $AP_s/(k_1\,N_E)$ becomes negligible with respect to one. Under these conditions it is possible to obtain a current proportional only to the external substrate concentration:

$$i = 2\,FA\,P_s\,c_s^{ex} \qquad (4)$$

With a given membrane of low permeability, the linear range of response increases with the amount of active enzyme. The use of a membrane of low permeability does not only extend the linear range of the calibration curve but also improves its time stability, because the decrease of the enzyme activity will have no influence on the current as long as $A.P_s$ is much smaller than $N_E.k_1$. It is evident that the lower limit of the membrane permeability is given by a tolerable proportion between the zero current of the sensor when lactate is absent and the signal current in the presence of lactate.

It should be mentioned that a lower membrane permeability does not necessarily lead to a longer response time. The solution of the time-dependent diffusion equation for one dimension shows that the response time of 95% of the steady state of the sensor system is a function of the membrane thickness, d, and the diffusion coefficient for lactate, D_S:

$$\tau_{95\%} = \frac{3,7\ d^2}{\pi^2\ D_S}, \qquad (5)$$

The relation between the permeability coeffizient, P_S, the diffusion coefficient, D_S, and the partition coefficient, $\mathcal{K}_S$ is given by:

$$P_S = \frac{\mathcal{K}_S \cdot D_S}{d} \qquad (6)$$

To achieve optimal conditions for linearity and drift behaviour the product of the permeability and the membrane area must be small. On the other hand, the diffusion coefficient should be high to achieve a short response time. With the proper choice of the membrane partition coefficient and the exposed area of the membrane both requirements can be met simultaneously.

To be of practical value, the lactate sensor should be usable for several weeks before enzyme and membrane renewal becomes necessary. The useful life of the sensor can be extended by working at low lactate levels. Therefore dilution of the sample by a factor of 10 is necessary.

The platinum anode of the sensor is cooled to 18^oC by a Peltier element, not only to prevent enzyme denaturation but also to standardise the temperature-dependent diffusion conditions.

The validity and precision of the method depends on the effect of a variety of substances found in blood or added to blood for several purposes, which may interfere with lactate measurements when this method is used. Interference could result from substances acting as substrates, activators, or inhibitors of the enzyme, or from substances that could be oxidised on the platinum anode and/or could reduce ferricyanide. To test the extent of the interference from the latter groups, 75 specimens of human blood plasma were collected and the lactate levels were determined with an operational device. In these samples the lactate concentration ranged from 0.7 to 10 mmol/ litre. The enzyme was then inactivated by heat denaturation and the background current of the sensor was determined with the same samples. The computed average value of the background current was found to be 88 nA, which corresponds to lactate concentrations of 0.1-0.2 mmol/ litre (3). These background errors can be considered to be negligible for most clinical applications.

In addition, the effect of a number of endogenous and/or related substances was investigated.

Particular attention was given to mono- and dicarboxylic α-hydroxy acids found in biological fluids that could be acted upon by the enzyme. Certain substances act as substrates for cytochrome b_2 and could therefore cause serious interference during lactate measurement if high concentrations were present in biological fluids. When the response to L-lactate is taken as 100%, the relative responses from these interfering substances are:

2-Hydroxybutyric acid	30%
2-Hydroxyhexanoic acid	30%
2-Hydroxy isocaproic acid	30%
Glyceric acid	30%
L-3-Phenyllactic acid	60%

Fortunately, in practice the concentrations of these substances in biological samples are sufficiently low for their interference to be negligible.

Interference from reducing substances can arise when such substances are directly oxidized at the anode or when they reduce ferricyanide. In the latter case the ferrocyanide produced will be oxidized, causing an excessively high reading.

Racine et al. (5) reported on the influence of such reducing substances.
Ascorbic acid (AA) gives negligible interference at normal concentrations. At millimolar concentrations substantial interference will occur, e.g., 0.18 mmol/litre AA increases the sensor reading by about

0.33 mmol lactate/litre and 1 mmol/litre AA by about 5 mmol lactate/ litre.

Uric acid increases the sensor current due to direct oxidation at the anode and reduction of ferricyanide. At concentrations four times higher than normal the lactate reading is 0.3 mmol/litre too high. An abnormally high uric acid concentration in the blood will cause elevated lactate readings.

The same authors published results on the influence of drugs commonly used in intensive care medicine. No interference from any of these drugs was detected. Most of the commonly used anticoagulants and preservatives do not interfere with the lactate assay (6). Only oxalate must not be used, since it is a very strong inhibitor and inactivates cytochrome b_2.

Extensive tests have established that lactate measurements in human plasma samples with the Lactate Analyzer are essentially the same as are obtained with the classic enzymatic methods based on the oxidation of lactate to pyruvate by NAD in the presence of lactate dehydrogenase. Linear regression analysis in 125 samples (cf. Fig. 2) indicates a one-to-one correspondence between the two methods and a correlation coefficient of nearly 1.

However, prior to the measurement the sample is diluted by a factor of 10; therefore, when comparing values in serum, plasma, and whole blood are compared in the same sample a difference related to the total cell mass can be expected and has been found (8).

The range of the device is 0-12 mmol/litre and it can be extended as needed by changing the dilution ratio. It is recommended that a 100-μl sample, which can be blood, serum, plasma, or amniotic, cerebrospinal, or extracellular fluid, is diluted. The volume of the cell is 50 μl. The precision (as standard deviation) is ±0.2 mmol/litre or 5% of the reading. The reproducibility (as the coefficient of variation) is ±2%. The response time is 40-70 s, giving a capacity of 20-30 measurements per hour. The useful life of the sensor is 3-4 weeks between renewals of the enzyme layer.

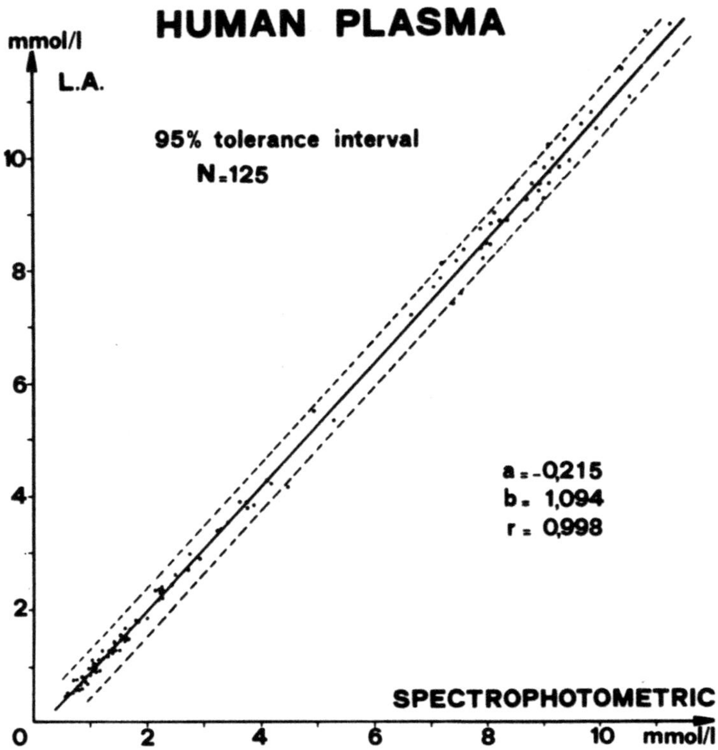

Fig. 2. Correlation of lactate measurement with the Lactate Analyzer and the spectrophotometric method. a = -0.215; b = 1.094; r = 0.998. Broken lines show 95% tolerance interval. A systematic difference is observed when whole blood is compared with plasma or serum samples (7). The classic method measures extra- and intracellular lactate, since the blood cells are destroyed during the addition of perchloric acid. The electrode, however, measures only the extracellular lactate and the measurement is practically independent of volume

Summary

The amperometric-enzymatic method of lactate analysis has been developed technically to a state where it can be considered a simple, fast, and reliable technique that gives results with adequate precision for most clinical applications.

References

1. WILLIAMS, D.L., DOIG, A.R., KOROSI, A.: Electrochemicalenzymatic analysis of blood glucose and lactate. Anal. Chem. 42, 118-121 (1970)
2. RACINE, P., MINDT, W.: On the role of substrate diffusion in enzyme electrodes. Biological Aspects of Electrochemistry, Experientia (Suppl.) 18, 525-534 (1971)
3. MINDT, W., RACINE, P., SCHLÄPFER, P.: Sensoren für Laktat und Glucose. Berichte der Bunsengesellschaft f. physikalische Chemie 77, 804-808 (1973)

4. RACINE, P., ENGELHARD, R., HIGELIN, J.C., MINDT, W.: An Instrument for the rapid determination of L-Lactate in biological fluids. Med. Instrum. 9, 11-14 (1975)
5. RACINE, P., KLENK, H.-O., KOCHSIEK, K.: Rapid lactate determination with an electrochemical enzymatic sensor: Clinical usability and comparative measurements. Z. Klin. Chem. Klin. Biochem. 13, 533-539 (1975)
6. GUARNACCIA, R., RACINE, P.: Investigation on the specificity of an amperometric enzymatic sensor for L-Lactate. Biomed. Techn. (Berlin) [Ergänzungsb.] 21, 189-190 (1976)
7. KRAGENINGS, I., RACKWITZ, R.: Bestimmung von Laktat nach enzymatisch-elektrochemischem Prinzip im Vergleich mit drei Modifikationen der enzymatischen Methode. Ärztl. Laboratorium 23, 549-554 (1977)
8. SOUTTER, W.P., SHARP, F., CLARK, D.M.: Bedside estimation of whole blood lactate. Br. J. Anaesth. 50, 445-450 (1978)

Closing Remarks

C.F. Rossel

First I should like to thank the speakers for their excellent papers.

One conclusion is obvious. The physician or investigator confronted with clinical, biological, or physiopathological problems will always have greater difficulty in grasping the theoretical side of the related sciences. A multidisciplinary approach is necessary, bringing together physicians, biologists, chemists, and physicists. This irreversible evolution will develop progressively as our knowledge increases, insofar as we really endeavour to adopt a scientific attitude.

Having said that, what is the impact of these papers at the medical level? The methods proposed by Boehringer and Hoffmann-La Roche differ relatively little as far as the principles are concerned. Both involve the enzymatic oxidation of lactate into pyruvate, with liberation of protons. The two methods therefore depend essentially, at this level on the "quality" (in the broad sense of the term) of the enzyme, on its inherent dynamism, and on its specificity. This enzymatic phase is followed by a physical measuring procedure, either amperometry or ultraviolet photometry.

In the case of electrochemical analysis there is simultaneous transformation of hexacyanoferrate into a reduced compound, which is reoxidised on a platinum electrode with the liberation of two electrons per lactate molecule; this gives rise to an electric current, which is measured by an ammeter.

In the case of spectrophotometric analysis, the oxidation of the lactate leads to the production of NADH, which absorbs the protons in a spectrophotometer. In this way the extinction is measured. This measurement phase is directly linked with the quality of the apparatus employed.

The paper presented by Professor Baudras clearly shows that within the limits of the concentrations encountered in medicine, the enzymatic transformations of lactic acid can be made stoichiometrically. It is also clear that the selectivity of the proposed enzymes is sufficient. In addition, the correlation between the spectrophotometric and the electrochemical methods is excellent, and falls within the limits of statistical error when human plasma is used. By contrast with whole blood, one must take into account the lactate concentration gradient between erythrocytes and plasma and build in a correction factor to compensate for the appearance of compartments linked to the interposition of the membrane at the level of the electrode. In a communication at the First World Congress on Intensive Care, London, 1975, I analysed this problem and demonstrated the difference (of illustration).

Thus, at our present state of knowledge, the two methods described by E. Munz and H.R. Wuhrmann are usable but not absolutely superimposable. For the sake of scientific accuracy, therefore, the method used should always be described.

Moreover, and on a more practical plane, one should not overlook the necessity for repeated calibration of the apparatus with fresh solutions. For a scientific investigation hourly calibration is not excessive, regardless of the method employed. In addition, one must remember that bacteria are likely to degrade the lactic acid; in consequence the calibration solutions should be stored in a refrigerator. Similarly, the production of lactic acid by the erythrocytes is considerable, but this is constantly neglected. This factor requires even more attention from the investigator than the apparatus, since the errors vary from case to case and are in no way repetitive.

Section III
Clinical Applications

Opening Remarks

C. Perret

I should like to begin by saying how pleased I am to attend this Symposium. I greatly appreciate the presence of experts from very different fields who are interested in lactate in one way or another. I feel that one of the advantages of a small meeting like this is precisely to give biochemists, physiologists, and clinicians the opportunity of exchanging ideas in a free discussion. Yesterday sessions were specially devoted to basic aspects of lactate metabolism. This constituted an excellent introduction to the clinical problems that will be presented this morning.

The most common cause of lactate accumulation in the tissues and blood is hypoxia. This includes the numerous circumstances during which there is an imbalance between the oxygen needs and the oxygen supply to the tissues. This type of hyperlactataemia was already recognised more than 60 years ago, but its pathogenesis is still not completely understood, and its clinical significance has yet to be specified. The most common cause of lactic acidosis is probably still muscular exercise, although it needs to be severe enough to induce a significant increase in the blood lactate concentration. Dr. Howald (Magglingen) will discuss this point.

Respiratory failure, whatever its cause, can be responsible for a severe decrease in arterial blood oxygen tension. Is this sufficient to impair the oxygenation of tissues, or are there adequate compensatory mechanisms in these situations? This is a question that Dr. de Coster and his colleagues will answer. The role of poor tissue perfusion in the pathogenesis of lactic acidosis will be discussed from the aspects of the effect of shock (Dr. Perret), heart disease (Dr. Messin), and peripheral arterial disease (Dr. Krahenbühl).

Overproduction of lactate is not necessarily related to hypoxia. Several conditions have been described in which hyperlactataemia develops in the absence of clinically detectable hypoperfusion or hypoxaemia. This particular variety of lactic acidosis has frequently been described in diabetes. Dr. Berger will discuss the different mechanisms involved in the development of lactic acidosis in diabetes and make some comments upon its clinical significance.

Chapter 12

Lactic Acid and Other Metabolic Indicators of Anaerobiosis in Normal Subjects

H. Howald

Introduction

Physical exercise is known to be a situation where even the normal subject may accumulate a high concentration of lactic acid in his working muscle cells, thus creating considerable metabolic acidosis of the blood and of the whole organism. But what type of physical activity is it that depends mainly on anaerobic processes? At what level of work intensity is more lactic acid produced than can immediately be metabolised? And does regular training influence lactic acid accumulation or blood acidosis tolerance?

Anaerobic and Aerobic Processes of Energy Supply

The metabolic processes involved in different types of physical activity and sports can be regarded as a function of work intensity and duration (Fig. 1). As a result of having available 5 μmol ATP and 15 μmol creatine phosphate per 1 g muscle tissue, human skeletal muscle can generate large amounts of work over a short period of time. Energy turnover is very high in sports activities such as weight-lifting or a 100-m sprint, and consequently cellular stores of energy-rich phosphates are almost depleted by these kinds of activity within approximately 10 s. But in spite of the involvement of large muscle masses, the general effect on the whole organism is minimal. Muscular work of up to 10 s duration does not depend on oxygen supply, nor does it produce measurable amounts of lactic acid (anaerobic, alactacid energy production).

For physical activity that must be maintained at a relatively high intensity for 2-3 min, ATP as a source of energy for muscular contraction is supplied by anaerobic glycolysis after the stores of

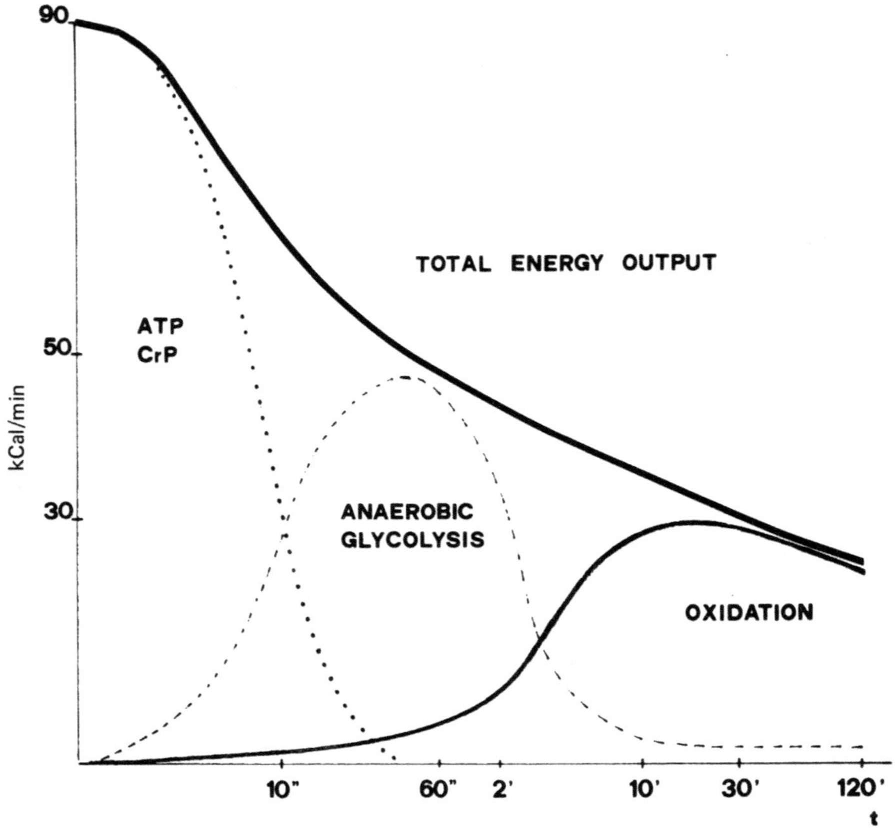

Fig. 1. Total energy output in trained human subjects in dependence on the duration of the activity (logarithmic scale on the abscissa). The time succession and quantitative amounts of energy produced by the different metabolic pathways are schematically represented. CrP, creatine phosphate

creatine phosphate are exhausted. The total energy turnover may be as high as 50 kcal/min for trained athletes such as 400-m and 800-m runners. High-intensity muscular work of 30s to 3 min duration produces large amounts of lactic acid as the end product of anaerobic glycolysis (anaerobic, lactacid energy supply). After exhaustive exercise, maximal lactate concentrations of 25-30 mmol/kg in the muscle and 20-25 mmol/litre in the blood have been measured in subjects engaged in running, swimming, speed skating, and rowing (1, 2). The level of acidosis resulting from these lactate concentrations is expressed in pH values of 6.3-6.4 in the muscle and 6.8-6.9 in arterial blood. These acidic conditions cannot be compensated by the blood buffers or by respiratory means, so that extreme acidosis is the reason for reducing work intensity or ceasing exercise, as a consequence of a metabolic feed-back inhibiting the activity of the glycolytic enzymes (3). During the recovery period following exhaustive activity the acid-base balance of the organism gradually becomes normal within 30-60 min. Lactic acid, which diffuses easily from the muscle into the blood stream, is subsequently metabolised by the myocardium and liver, and by the skeletal muscle itself. If light work is performed during the recovery phase the acidic conditions are reversed much more quickly (4).

During muscular work of more than 3 min duration, ATP is supplied mainly by oxidative metabolism, even though there may be some overlap with lactic acid production in activities lasting up to 10 min. Well-trained endurance athletes, such as cross-country skiers, cyclists, and long-distance runners, can have an energy output of approximately 30 kcal/min from aerobic sources. However, such a high rate cannot be maintained indefinitely, and during such events as the marathon, for example, energy output is usually only 75%-80% of the maximum aerobic capacity. The blood lactic acid concentration after a long distance run of more than 30 min rarely exceeds 8-10 mmol/litre and in events of even longer duration lactic acid concentrations are usually at a level only two to three times the resting values (5).

The Anaerobic Threshold

Performance capacity in physical activities lasting more than 2-3 min is limited by an individual's maximum oxygen uptake capacity ($\dot{V}_{O_2}$ max.), which varies between 45 and 50 ml/min · kg body weight for sedentary men aged 20-30 years and 80-85 ml/min · kg for well-trained cyclists, long-distance runners, and cross-country skiers. $\dot{V}_{O_2}$ max. can only be attained at heavy work intensities also leading to metabolic acidosis, because of an imbalance between lactic acid production in the muscles and the body's capacity for its uptake or removal. Lactate production is then so high that the work intensity eliciting the maximum aerobic capacity can be sustained only for a very few minutes, while the pace has to be reduced to about 90%-95% of $\dot{V}_{O_2}$ max. in a 5000-m, 85%-90% in a 10000-m, and 75%-80% in a marathon run. The work intensity at which blood lactic acid concentrations begin to rise significantly above the resting values has been termed the anaerobic threshold (6), which can be expressed as a percentage of a subject's $\dot{V}_{O_2}$ max. Recently it has become apparent that besides $\dot{V}_{O_2}$ max., the level of an individual's anaerobic threshold is a critical parameter in determining his capacity for prolonged physical exercise.

The anaerobic threshold can be determined by either bicycle ergometry or treadmill running in a test protocol with progressively increasing workloads (7). From respiratory data it is evident that at a certain percentage of $\dot{V}_{O_2}$ max., both ventilation and the amount of expired CO_2 begin to increase at a steeper rate than they do at lower levels of work intensity and oxygen consumption (Fig. 2). The respiratory equivalent ($\dot{V}_E/\dot{V}_{O_2}$), expressing the economy of ventilation, loses its linear relationship to increasing work loads at the same percentage of $\dot{V}_{O_2}$ max. The sudden changes in the respiratory pattern are a direct consequence of the exercise-induced increases in blood lactic

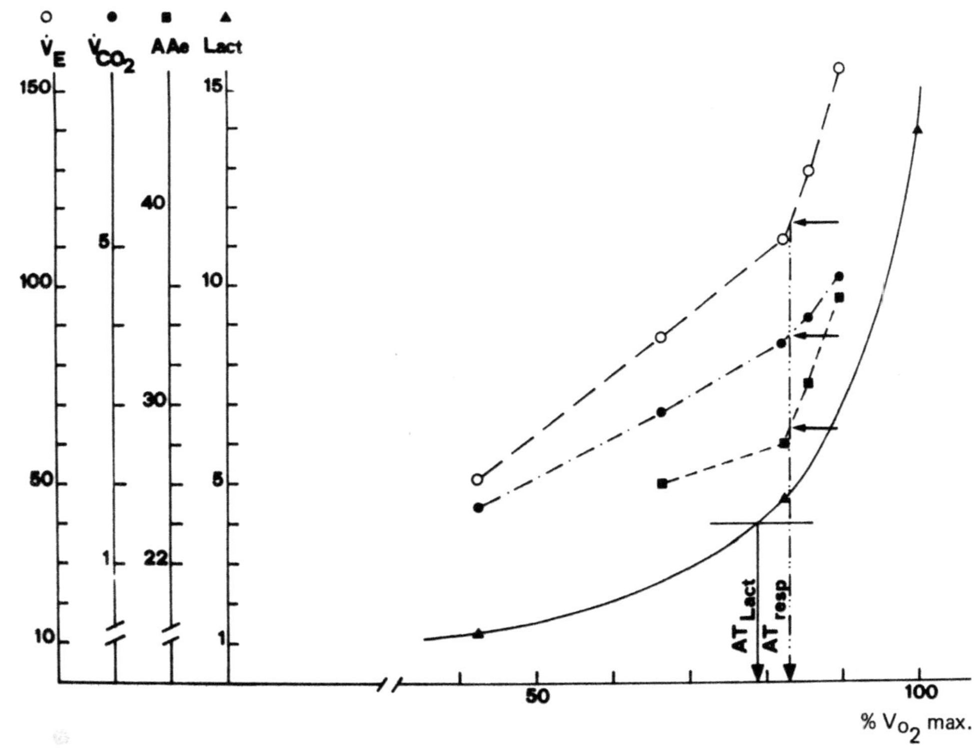

Fig. 2. Determination of the anaerobic threshold from respiratory data (AT_{resp}) or blood lactic acid concentration (AT_{Lact}). Arrows indicate the sharp increase of ventilation ($\dot{V}_E$; litres/min), carbon dioxide excretion ($\dot{V}_{CO2}$; litres/min), and respiratory equivalent (AAe) at a work intensity leading to an oxygen uptake of 83% of maximum. AT_{Lact} is derived from the exponentially increasing blood lactic acid concentrations at a critical level of 4 mmol/litre

acid concentration. The critical level of these has been empirically found to be 4 mmol/litre by different authors (8, 9).

Our own findings, shown in Fig. 2, suggest to us that the concentration of 4 mmol/litre seems to characterise the point where respiratory compensation of metabolic acidosis begins to play an important role. Determination of this critical point by measurement of the blood lactic acid concentration is much easier than the spirometric approach reserved to well-equipped laboratories. We found a good correlation between the anaerobic thresholds evaluated by either respiratory data or the 4 mmol/litre blood lactic acid level when testing 35 athletes involved in ice hockey, rowing, cycling, and long-distance running (Fig. 3). The highest values were observed in the long-distance runners (AT_{Lact} = 92.5%±4.0% of $\dot{V}_{O_2}$ max.), the lowest ones in the ice hockey players (77.6%±3.0% of $\dot{V}_{O_2}$ max.). Sedentary people and cardiac patients are reported to show even lower values of about 50%-60% of $\dot{V}_{O_2}$ max. (10).

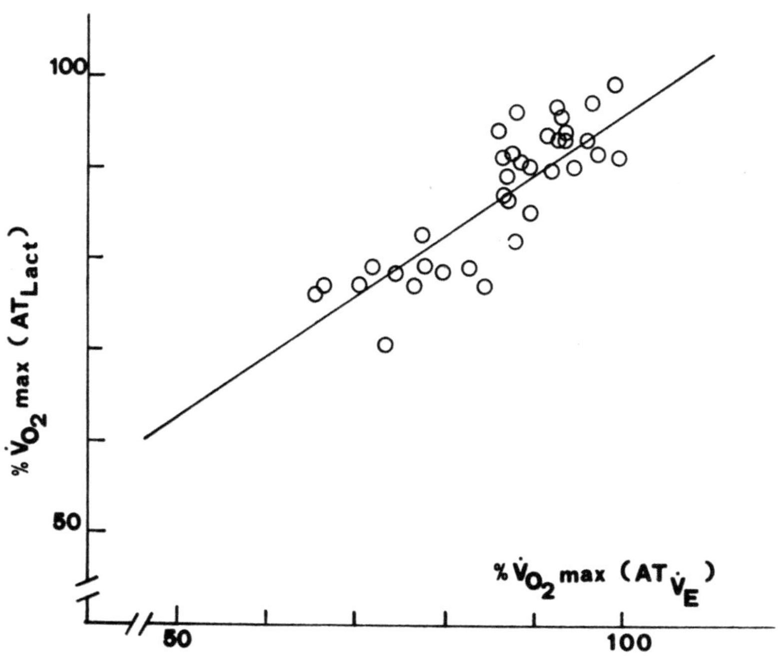

Fig. 3. Correlation between anaerobic threshold (AT) determined from respiratory data and blood lactic acid concentrations recorded in 35 subjects. y = 29.92 + 9661 x; r = 0.85

There are two main forms of physical training in which the lactic acid concentration in both muscle cells and the whole organism is of major importance: continuous prolonged work (endurance training) and repeated short bursts of activity separated by short periods of rest (intermittent exercise or interval training).

Endurance training is aimed at an improvement of a subject's aerobic capacity ($\dot{V}_{O_2}$ max.) on the one hand, and in the development of a high anaerobic threshold on the other. The best training stimulus seems to come from prolonged muscular work maximally involving the oxidative processes of the cellular metabolism. The exercise-induced adaptation then consists in an increase in the mitochondrial volume (11) and in the activities of oxidative enzymes in the muscle cells (12, 13). Lactic acid accumulation is believed to disturb these adaptive processes to a certain degree, and thus the optimum training intensity should not exceed the anaerobic threshold. Obviously, repeated measurements of blood lactic acid concentrations cannot be performed

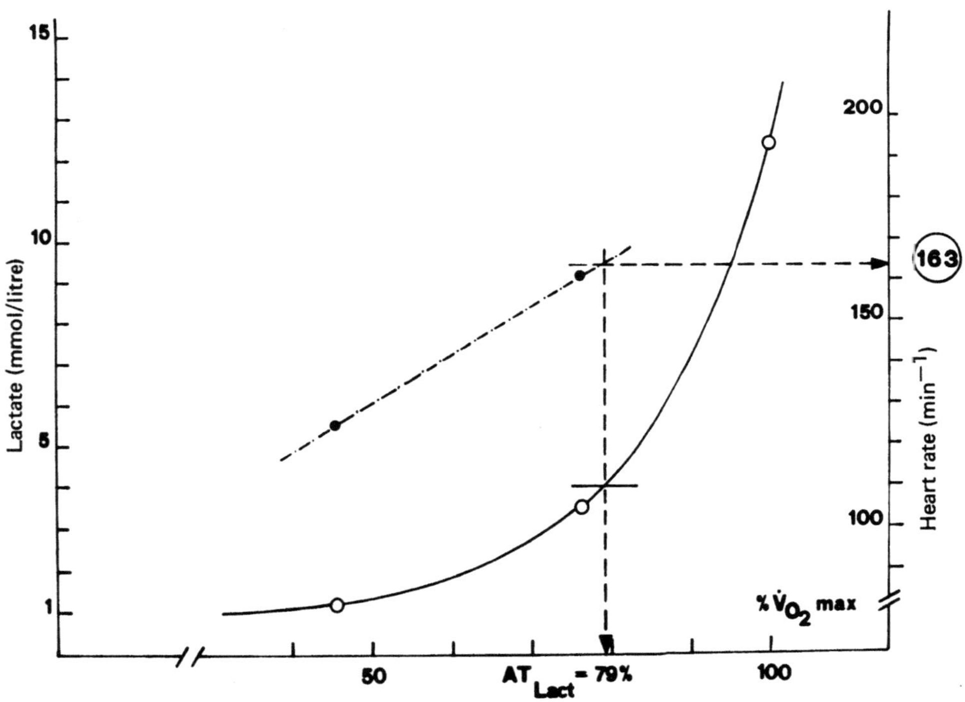

Fig. 4. Determination of the critical heart rate for training at the anaerobic threshold of 4 mmol/litre blood lactic acid concentration (AT_{Lact})

in the daily training routine. But the linear increase of heart rate with increasing work intensity under steady-state conditions helps to overcome this difficulty. The steady-state heart rate related to a workload bringing the blood lactic acid concentration to the critical level of 4 mmol/litre can be extrapolated according to Fig. 4. The athlete or the patient should then work at an intensity eliciting nearly this critical heart rate but never exceeding it, to achieve an optimum training effect for endurance. The anaerobic threshold being lower in patients and sedentary people than in trained athletes, the critical heart rate will also be at a lower level for them, at least at the beginning of a training programme.

Interval training stimulates mainly the alactacid and lactacid mechanisms of anaerobic energy production, depending on the duration of the single bouts of heavy exercise on one hand and on that of the rest periods between these working phases on the other. Severe local and general acidosis is generated by muscular work at 90%-95% of an individual's maximum capacity for 30-120 s followed by 2-10 min rest, repeated several times. In spite of blood lactic acid concentrations as high as 10-15 mmol/litre at the end of a resting interval, the muscles and the organism are well able to complete the following working period at the same intensity they had maintained during the preceding ones. An interval training programme enhances the activities of cellular enzymes involved in the anaerobic glycolysis, such as phosphofructokinase and glyceraldehyde 3-Phosphate dehydrogenase (14). Moreover, the athlete specifically trained for anaerobic work is able to tolerate much higher blood lactic acid concentrations and lower blood pH values during exhaustive exercise than an untrained person. According to these observations, the effect of anaerobic training seems to be a faster turnover rate of the cellular anaerobic glycclysis in muscle and a general adaptation to acidosis.

Summary

Peak blood lactic acid concentrations up to 20 mmol/litre are frequently observed in athletes competing in events such as 400-800 m running, 100-200 m swimming, 400 m speed skating, and rowing, as a consequence of high-intensity muscular work over a relatively short time (a few minutes). Respiratory compensation of exercise-induced metabolic acidosis starts at a blood lactic acid concentration of approximately 4 mmol/litre, and this critical point is termed the "anaerobic threshold". Depending on the grade of physical fitness, the anaerobic threshold is reached at a workload eliciting 60%-90% of an individual's maximum oxygen uptake capacity or a heart rate of 140-170/min. The optimum endurance training effect can be attained with continuous work almost at the anaerobic threshold for a minimum of 15-20 min. Interval training with repeated short bouts of exercise at near-maximum intensity separated by short resting periods can improve both the cellular anaerobic capacity of the muscles and the tolerance of acidosis in the whole organism.

References

1. HERMANSEN, L., OSNES, J.B.: Blood and muscle pH after maximal exercise in man. J. Appl. Physiol. 32, 304-308 (1972)
2. KINDERMANN, W., KEUL, J.: Anaerobe Energiebereitstellung im Hochleistungssport. Schorndorf: Hofmann 1977
3. McGILVERY, R.W.: The use of fuels for muscular work. In: Metabolic adaptation to prolonged physical exercise. Howald, H., Poortmans, J.R. (eds.), pp. 12-30. Basel: Birkhäuser 1975
4. HERMANSEN, L., MAEHLUM, S., PRUETT, E.D.R., VAAGE, O., WALDUM, H., WESSEL-AAS, T.: Lactate removal at rest and during exercise. In: Metabolic adaptation to prolonged physical exercise. Howald, H., Poortmans, J.R. (eds.), pp. 101-105. Basel: Birkhäuser 1975
5. ASTRAND, P.O., HALLBÄCK, I., HEDMAN, R., SALTIN, B.: Blood lactates after prolonged severe exercise. J. Appl. Physiol. 18, 619-622 (1963)
6. WASSERMANN, K., WHIPP, B.J., KOVAL, S.N., BEAVER, W.I.: Anaerobic threshold and respiratory gas exchange during exercise. J. Appl. Physiol. 35, 236-243 (1973)
7. SCHÖNHOLZER, G., BIELER, G., HOWALD, H.: Ergometrische Methoden zur Messung der aeroben und anaeroben Kapazität. In: 3. Int. Seminar für Ergometrie. Hansen, G., Mellerowicz, H. (eds.), Berlin: Econ 1972
8. KINDERMANN, W., SIMON, G., KEUL, J.: Dauertraining; Ermittlung der optimalen Trainingsherzfrequenz und Leistungsfähigkeit. Leistungssport 8, 34-39 (1978)
9. MADER, A., LIESEN, H., HECK, H., PHILIPPI, H., ROST, R., SCHÜRCH, P., HOLLMANN, W.: Zur Beurteilung der sportartspezifischen Ausdauerleistungsfähigkeit im Labor. Sportarzt Sportmed. 27, 80-88 und 109-112 (1976)
10. WASSERMAN, K., WHIPP, B.J.: Exercise physiology in health and disease. Am. Rev. Respir. Dis. 112, 219-249 (1975)
11. HOPPELER, H., LÜTHI, P., CLAASSEN, H., WEIBEL, E.R., HOWALD, H.: The ultrastructure of the normal human skeletal muscle; a morphometric analysis on untrained men, women and well-trained orienteers. Pfluegers Arch. 344, 217-232 (1973)
12. HOWALD, H.: Ultrastructure and biochemical function of skeletal muscle in twins. Ann. Hum. Biol. 3, 455-462 (1976)
13. MOESCH, H., HOWALD, H.: Hexokinase (HK), Glyceraldehyde-3P-Dehydrogenase (GAPDH), Succinate-Dehydrogenase (SDH) and 3-Hydroxyacly-CoA-Dehydrogenase (HAD) in skeletal muscle of trained and untrained men. In: Metabolic adaptation to prolonged physical exercise. Howald, H., Poortmans, J.P. (eds.), pp. 463-465. Basel: Birkhäuser 1975
14. PFISTER, B., MOESCH, H., HOWALD, H.: Beeinflussung glykolytischer und oxidativer Skelettmuskelenzyme des Menschen durch anaerobes Training oder anabole Steroide. Med. H. Sport, to be published

Chapter 13

Lactate and Pulmonary Pathology

A. de Coster, R. Messin, and S. Degré

In beview of the well-known effect of hypoxia on cellular and anaerobic metabolism, there must be an interesting relationship between lactate (LA) and chronic pulmonary disease.

Materials and Methods

All the patients involved in our study were suffering from a chronic bronchopulmonary disease or from an acute attack after previous bronchopulmonary damage. Blood gases were measured with classic membrane electrodes.

Lactate (LA) and pyruvate (PY) were determined by the enzymatic method and measured in plasma, the concentrations of both LA and PY in plasma always being higher than the whole blood concentrations. This procedure allows detection of sometimes slight and transient increases in LA or PY, which might perhaps be "masked" if measurements were made in whole blood.

Regression lines for concentrations in red cells and plasma have been established previously (1).

Production or Removal of LA by the Lungs?

Controversial results have been proposed from a physiological point of view. It has been well demonstrated by Tierney (2) that lung slices can produce a significant amount of LA in vitro and that glucose may disappear. No, such observations are obvious in vivo, as illustrated in Table 1. We have measured LA in arterial (LAa) and mixed venous (LA $\bar{v}$) blood in a wide variety of patients with bronchopulmonary disease. Our results show clearly that if the blood gases

Table 1. LA in arterial blood (LAa) and in mixed venous blood (LA $\bar{v}$) of patients with bronchopulmonary disease

	n	LAa mg%	LA $\bar{v}$ mg%	Arterio venous difference	
PaO_2>75 - $PaCO_2$ N	17	10.4 ± 4.5	10.3 ± 5.1	0.06 ± 3.1	NS
PaO_2<75 - $PaCO_2$ N	23	9.2 ± 3.5	8.7 ± 3.7	0.46 ± 2.3	NS
PaO_2<75 - $PaCO_2$>45	30	10.1 ± 4.7	10.5 ± 4.6	0.4 ± 1.7	NS

were normal or if there was hypoxia (whether or not combined with hypercapnia), there were no arteriovenous differences in LA. These observations are in agreement with those of Rochester and co-workers (3), although these authors have demonstrated that samples taken from a vein draining a lobe with a malignant process or active tuberculosis may have a higher LA level; even in these conditions, however, the arteriovenous difference is undetectable, the small amount of LA produced being diluted by the blood stream.

Our results are also in agreement with those of Mitchell et al. (4) and Harris et al. (5). We have never observed cases where the lung could produce LA or remove LA from the blood stream (6).

Relationship Between LA and Blood Gases in Chronic Obstructive Pulmonary Disease

At rest, there is no correlation between the degree of hypoxia and LA. For a wide range of PaO_2, LA remains at around 10 mg%, although some higher values may be observed in hypoxic and in normal patients. Such high "normal resting levels" are probably related to some muscular contractions or psychotic hyperventilation.

In the same group of patients, we induced hypoxia by reducing the FiO_2 to levels between 10% and 14%. Sometimes very significant hypoxaemia, with a decrease in PaO_2 to 30-40 mmHg, were observed. Blood samples were taken at the end of this induced hypoxaemia.

Even a very significant fall in PaO_2 does not necessarily induce any significant rise in LA. On the other hand, in the same group of patients the administration of pure oxygen for a short time raised the PaO_2 to high levels without any significant modification of LA level.

In the same way, we did not find any correlation of variation of $PaCO_2$ with high or normal LA level. Our results are somewhat different from those of Biot et al. (7), who observed a tendency for higher levels of LA in arterial hypoxaemia, but these authors also stated that the increase in LA, if it exists, does not exceed 20 mg/ml.

Lactate and Asthmatic Attacks

In most papers dealing with acute attacks of asthma, respiratory alkalosis and moderate hypoxaemia are reported in most cases. In very severe attacks, which may induce a true comatose state, much more severe and less typical blood gas disturbances have been described.

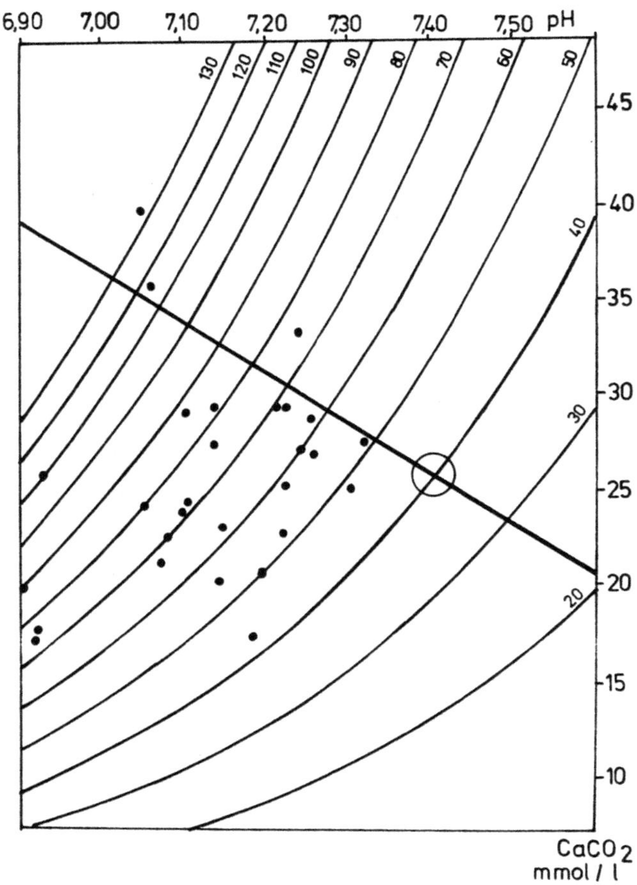

Fig. 1. Values recorded for pH and $CaCO_2$ (mmol/litre) in comatose asthmatic patients

They are consistent with very severe respiratory insufficiency with hypoxaemia and hypercapnia.

We have had the opportunity of observing several cases of comatose patients with very severe bronchial obstruction. More than 50% of them had typical atopic asthma and were admitted with very severe hypoxaemia and hypercapnia (Fig. 1). In some cases, very low values of pH (below 7) are attributable to shock or cardiac arrest, but in others these abnormalities are not involved although the haemodynamic situation is frequently characterised by acute pulmonary hypertension, as may be assumed by very high and transitory pulmonary P waves in the ECG.

In our cases, there is obviously a combination of metabolic and respiratory acidosis, the respective part of each varying in different patients. In a few of these patients in whom it was measured, LA was high.

Metabolic acidosis in acute asthmatic attacks was also observed by Rees et al. (8), Holmdahl et al. (9), Lissac et al. (10), and Mithoefer et al. (11), but this acidosis was always slight and occurred in only a few patients in the series studied. Roncoroni (12), in his series of 103 cases of acute attacks of asthma, found mixed acidosis in only 14 patients the metabolic component being attributed to an increase in LA.

Acute hypoxaemia may not be responsible for this increase in LA, as we have demonstrated before in acute hypoxic experiments. Alkalosis can also be excluded in our group, as can starvation; in some cases, slight aceton disappears rapidly from the urine with glucose infusion. The mechanism of the development of metabolic acidosis in status asthmaticus is complex: previous alkalosis before the comatose state; increased work of respiratory muscles; poor tissue perfusion; decrease in LA uptake by the liver; decrease in cardiac output; and increase in circulating catecholamines may all be contributing factors.

Lactate During Exercise in Patients with Chronic Obstructive Pulmonary Disease

In a large sample of patients with chronic pulmonary disease, exercise was performed on a bicycle over 15 min with a constant load. Due to the severity of the illness, the load does not exceed 40 W or, in a few cases, 70 W. In all these cases, blood samples were taken after 3, 5, 10 and 15 min of exercise. The level of LA was high compared with the $\dot{V}_{O_2}$ in all the patients studied. In normal subjects at the same level of exercise, LA always increases, but generally the increase is not so significant and is transitory, the level being normal at the end of the physical work. So, in normal subjects, a transient increase of LA may only be objectivised if the samples are numerous and are collected at the onset of exercise.

We have tried to find correlations between the significant increase of LA in patients with bronchopulmonary obstructive disease and other parameters, but all our attempts have failed.

A decrease, sometimes significant, of PaO_2 does not imply a higher level of LA. The degree of hyperventilation or hypercapnia, the cardiac output, whose evolution is normal during physical work, and the pulmonary hypertension show no relationship with the significant increase in LA. Our observations are also in agreement with those of Gimenez (13).

The only fact that we have to keep in mind is that high levels of LA seem to be correlated with high levels of heart rate and low stroke volume in patients with bronchopulmonary disease, as in normal subjects. In a few patients, the evolution of the LA:PY ratio was studied several times during exercise.

As reported previously by others and ourselves, the evolution of LA and PY is not identical. The increase in LA is earlier, with a later increase of PY, so that the LA:PY ratio is higher at the beginning of exercise. However, if the evolution of the LA is reported to follow three or four different patterns (14), the evolution of PA seems much more variable, as was also stated by Segall (15).

Lactic Acid and Pulmonary Circulation

The role of the hydrogen ion concentration on the pulmonary vascular bed has already been demonstrated by Bergovsky et al. (16), Enson et al. (17), and Harvey (18). Gimenez (13) also found a good correlation between the mean pulmonary arterial pressure and the lactate level of arterial blood in bronchitic patients (9 patients), while in other patients with restrictive disease higher levels of LA do not seem to be correlated with higher pulmonary arterial pressure.

We think that it would be better to correlate LA with pulmonary arterial resistance, which in fact controls the level of pulmonary arterial pressure for a given cardiac output and a given level of right ventricular work. There is some correlation between LA and total pulmonary vascular resistance. The same conclusion arises if we compare pH or hydrogen ion concentrations and cardiac output and pulmonary arterial pressure.

Of course, in patients with severe bronchopulmonary disease, during exercise we will observe a fall in pH and PaO_2 and an increase in $PaCO_2$, LA, and mean pulmonary arterial pressure, but this does not imply that all these factors are necessarily correlated by a cause-effect mechanism.

Lactic Acid and Rehabilitation of Patients with Chronic Obstructive Pulmonary Disease

It is well known that LA is lower in athletes and after training, because of a better adaptation of the local circulation of working muscles.

Nine patients with severe obstructive pulmonary disease took part in a training programme (bicycling at 60%-80% of the maximal oxygen consumption SL (Symptom limited) during 15-20 min three times weekly). These patients were also trained to use diaphragmatic breathing with a lower frequency and a higher tidal volume. They were studied before and after training, and the only modifications we observed were a slight decrease of pulmonary hypertension and a slight improvement of PaO_2, the level of LA remaining constant for the same amount of work (19).

Conclusions

Our investigations on LA in bronchopulmonary disease have yielded disappointing results. In fact, the disturbance of the mechanics of the lung may induce increased work of breathing, acid-base disturbances, hypoxaemia, and pulmonary hypertension, but these factors are not correlated with the level of LA. The cardiac output is not involved; in the cases studied it was normal both at rest and during exercise. Thus we have the impression that the most important factor is a peripheral one related to oxygen transport to the working muscles. At the beginning of physical exercise, the LA produced will probably be less if oxygen transport to the working muscles is adapted to the energetic expenses as soon as possible. Unfortunately, it is not easy to follow this phenomenon, which may be most important in the first seconds immediately after the beginning of the exercise.

PVO_2 may perhaps be an interesting parameter, but the LA produced by working muscles is diluted in the total blood volume. So perhaps it would be more interesting to monitor the PO_2 of the blood from a working leg, or better, the PO_2 coming from a working muscle, at different points during exercise. This phenomenon must be the main one involved in LA production in all of our patients who are untrained.

References

1. DE COSTER, A., MESSIN, R., FRANCKSON, J.R.M.: Répartition de l'acide lactique entre plasma et globules rouges chez l'homme. Comparaison des résultats fournis par différentes méthodes de dosage. Arch. Int. Physiol. Biochim. 73, 251-258 (1966)
2. TIERNEY, D.F.: Intermediary metabolism of the lung. Fed. Proc. 33, 2233-2237 (1974)
3. ROCHESTER, D.F., WICHERN, W.A., FRITTS, H.W., Jr., CALDWELL, P.R.B., LEWIS, M.L., GIUNTINI, C., GARFIELD, J.W.: Arteriovenous differences of lactate and pyruvate across healthy and diseased human lung. Am. Rev. Respir. Dis. 107, 442-448 (1973)
4. TIERNEY, D.F., LEVY, S.E.: Glucose metabolism. In: The Biochemical basis of pulmonary function. Crystal, R.G. (ed.), pp. 105-125. New York, Basel. Dekker 1976
5. HARRIS, P., BAILEY, T., BATEMAN, M., FITZGERALD, M.G., GLOSTER, J., HARRIS, E.A., DONALD, K.W.: Lactate, pyruvate, glucose and free fatty acids in mixed venous and arterial blood. J. Appl. Physiol. 18, 933 (1963)
6. HALMAGYI, D., REINBOLD, A., FELKAI, M., YVANYI, J.: Zur Frage der intrapulmonalen Milchsäureoxydation beim Menschen. Klin. Wochenschr. 29, 219 (1951)
7. BIOT, N., KOFMAN, J., CASSAN, G., PERRIN-FAYOLLE, M., BRUN, J.: La lactacidémie au cours des insuffisances respiratoires chroniques: ses variations au repos et à l'effort et ses corrélations avec l'hypertension pulmonaire. Poumon Coeur 27, 555-574 (1971)

8. REES, H.A., MILLAR, J.S., DONALD, K.W.: A study of the clinical course and arterial blood gas tensions of patients in status asthmaticus. Q. J. Med. 37, 541-561 (1968)
9. HOLMDAHL, M.H., HEDSTRAND, U., PARROW, A., KORKEILA, J., TELIVUO, L., MATELL, G.: Association respiration artificielle-THAM au cours du traitement de l'état de mal asthmatique. Presse Méd. 75, 957-960 (1967)
10. LISSAC, J., LABROUSSE, J., TENAILLON, A., BOUSSER, J.P., LABROUSSE, F., JACQUOT, Ch.: Les désordres acido-basiques dans l'état de mal asthmatique. Bull. Eur. Physiopathol. Respir. 11, 745-756 (1975)
11. MITHOEFER, J.C., PORTER, W.F., KARETZKY, M.S.: Indication for the use of sodium bicarbonate in the treatment of intractable asthma. Respiration 25, 201-215 (1968)
12. RONCORONI, A.J., ADROGUE, H.J.A., DE OBRUTSKY, C.W., MARCHISIO, M.L., HERRERA, M.R.: Metabolic acidosis in status asthmaticus. Respiration 33, 85-94 (1976)
13. GIMENEZ, M., ROMERO COLOMER, P., HENNEQUIN, R., SAUNIER, C.: Hyperlactacidémie et hémodynamique pulmonaire à l'exercise modéré chez les insuffisants respiratoires chroniques. Bull. Eur. Physiopathol. Respir. 10, 281-300 (1974)
14. DE COSTER, A., MESSIN, R., DEGRE, S.: Blood changes in lactic and pyruvic acid during exercise. In: Ergebnisse der Ergometrie, Mellerowicz, H., Jokl, E., Hansen, G. (eds.), pp. 207-210. Erlangen: Straube 1975
15. SEGALL, D.: Effects of exercise in lung disease: the lactate response. Am. J. Med. Sci. 262, 218-226 (1971)
16. BERGOFSKY, E.H., LEHR, D.E., FISHMAN, A.P.: The effect of changes in hydrogen ion concentration on the pulmonary circulation. J. Clin. Invest. 41, 1492-1502 (1962)
17. ENSON, Y., GIUNTINI, C., LEWIS, M.L., MORRIS, Q.T., FERRER, I., HARVEY, R.M.: The influence of hydrogen ion concentration and hypoxia on the pulmonary circulation. J. Clin. Invest. 50, 2703-2714 (1964)
18. HARVEY, R.M.: The influence of hydrogen ion in the control of pulmonary artery pressure in patients with obstructive disease of the lungs. Med. Thor. 22, 108-117 (1965)
19. DEGRE, S., SERGYSELS, R., MESSIN, R., VANDERMOTEN, P., SALHADIN, P., DENOLIN, H., DE COSTER, A.: Hemodynamic responses to physical training in patients with chronic lung disease. Am. Rev. Respir. Dis. 110, 395-402 (1974)

Chapter 14

Lactic Acid in Shock and Liver Failure

C. Perret and J.F.Enrico

Introduction

Metabolic acidosis during prolonged shock is a well-recognised and documented fact. As shown by Cournand et al. (1) in 1943, acidosis is linked with lactate accumulation in extracellular fluid, which is due to an imbalance between tissue oxygen supply and metabolic needs.

In 1964, Peretz et al. (2) demonstrated that the blood lactate level was a useful prognostic index in shock. The same year, Broder and Weil (3) reported similar results with an excess lactate level, which, according to Huckabee (4), was thought to be a reliable index of oxygen debt. They reported that the prognosis is very poor once the excess lactate level exceeds 4 mmol/litre. So, by the simple determination of lactate level, it seemed the clinician could judge the severity of shock and predict the outcome. However, many objections were raised to this point of view, especially in septic shock, where no correlation was observed between measured oxygen deficit and lactate levels. In these conditions, lactataemia appeared to be a poor guide to the extent of cellular anoxia (5).

The present study was undertaken to evaluate the acid-base status in shock, to test the prognostic value of blood lactate concentration, and finally to determine whether other factors than anoxia were likely to interfere with lactate metabolism.

A group of 84 patients was studied, made up of 31 cases of hypovolaemic shock, 25 cases of cardiogenic shock, and 28 cases of septic shock.

A diagnosis of shock was established in the presence of the following criteria: systolic arterial pressure below 90 mmHg, a urine output of less than 25 ml/h associated with peripheral sweating, cyanosis of the extremities, and possible mental confusion.

Arterial blood samples were collected in all the patients upon admission and regularly throughout the evolution.

Acid-base determinations were performed in arterial blood by the Astrup whole-blood microequilibration technique. Lactate and pyruvate were determined by the enzymatic method described by Bucher. The normal lactate value for our laboratory is 0.97 mmol/litre (± 0.08) and the normal pyruvate value 0.097 mmol/litre (± 0.008). These values correspond to a normal L:P ratio of 10.

Results

Individual values obtained upon admission for hydrogen ion concentration and PCO_2 are presented in Fig. 1, where PCO_2 (mmHg) is plotted along the abscissa and the hydrogen ion concentration (nEq/

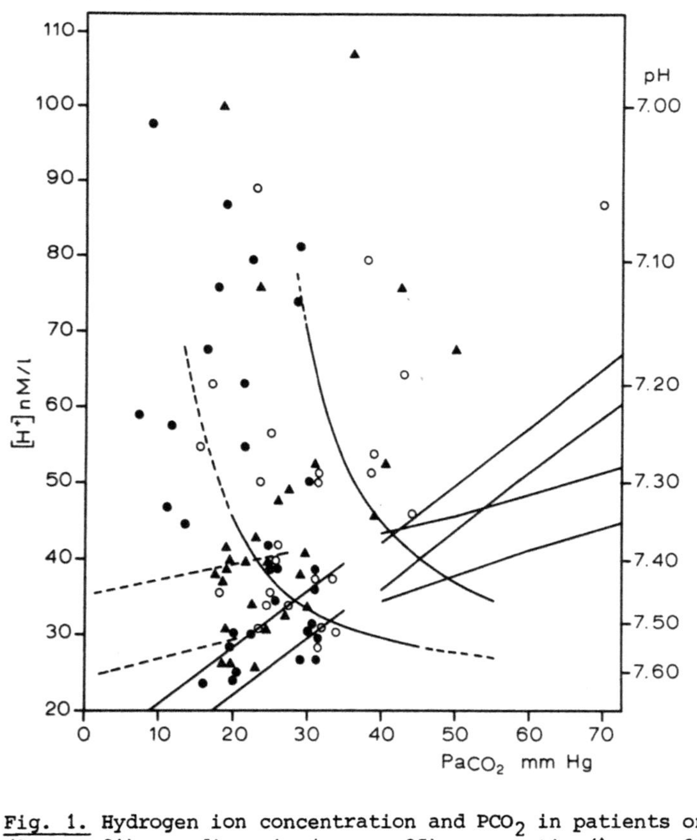

Fig. 1. Hydrogen ion concentration and PCO_2 in patients on admission in hypovolaemic (●; n = 31), cardiogenic (o; n = 25), or septic (▲; n = 28) shock

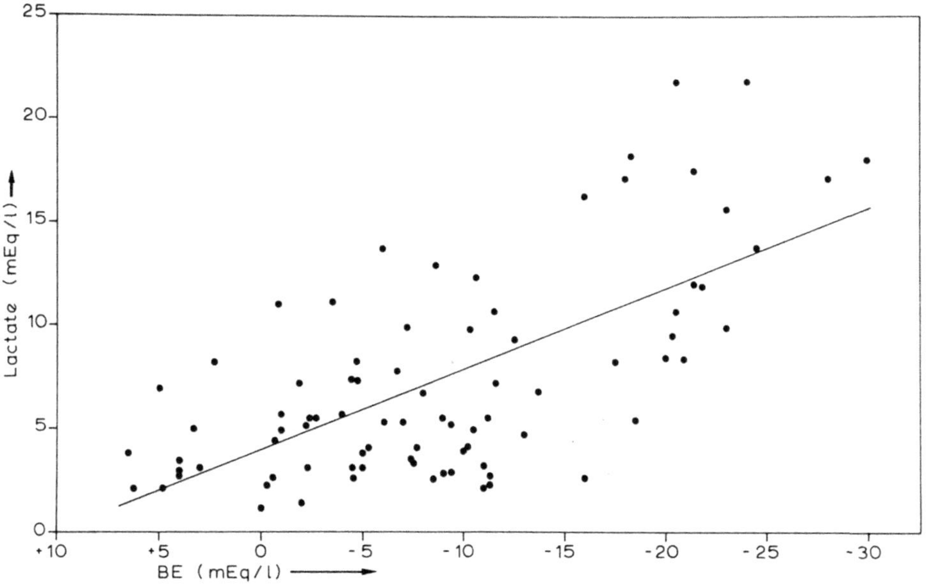

Fig. 2. Arterial base excess BE and lactate concentration in patients in shock on admission ($\underline{n}$ = 84). y, 3.98-0.39x; $\underline{r}$, 0.645; $\underline{P} < 0.001$

litre) along the ordinate. The dots designate cases of haemorrhagic shock, open circles cases of cardiogenic shock, and triangles cases of septic shock.

It is clearly apparent that there is no evident relation between the aetiology of shock and the type of acid-base disorder observed. The most common disorder observed in our series upon admission was an acute or chronic respiratory alkalosis (43 patients). In 24 cases there was a pure metabolic acidosis. In 17 cases a mixed disorder was observed. The correlation between negative base excess (or base deficit) and lactate concentration is significant but the coefficient of correlation is low ($\underline{r}$ = 0.65) (Fig. 2). Some patients have a high lactate value with a positive base excess. Others have significant base deficit without a corresponding lactate level. This discrepancy probably has many causes, including the presence of acid-base disorders preceding the development of shock and the release of other acids than lactic acid.

Thus it may be concluded that the simple determination of base deficit cannot be substituted for measurement of the arterial lactate concentration. A correlation between lactate and excess lactate was established in a series of 115 patients, made up of 84 shock patients and 31 cases of cardiac arrest, who were successfully resuscitated. The correlation coefficient is 0.97. It shows that lactate and excess lactate concentrations correspond very closely. Some excess lactate values will be presented in later figures to demonstrate that there is no particular benefit in using this concept, which has been widely adopted despite major criticisms concerning its significance.

To estimate the prognostic value of arterial blood lactate, the relation between the mortality rate and the maximal blood lactate level measured during evolution was established (Fig. 3). Patients were divided into seven groups according to their blood lactate concentration. The number above each column represents the number of patients in each group. On the whole, it is evident that the survival rate decreases with the increase of lactate levels. However, two facts should be pointed out. On the one hand, some patients had a fatal outcome although blood lactate was only moderately increased. On the other hand, four patients survived with lactate blood levels as high as 13 or 18 mEq/litre. Such a discrepancy suggests that lactataemia does not depend only on the intensity and duration of tissue anoxia. Other factors might be implicated, which could interfere with lactate metabolism and in this way diminish its prognostic value.

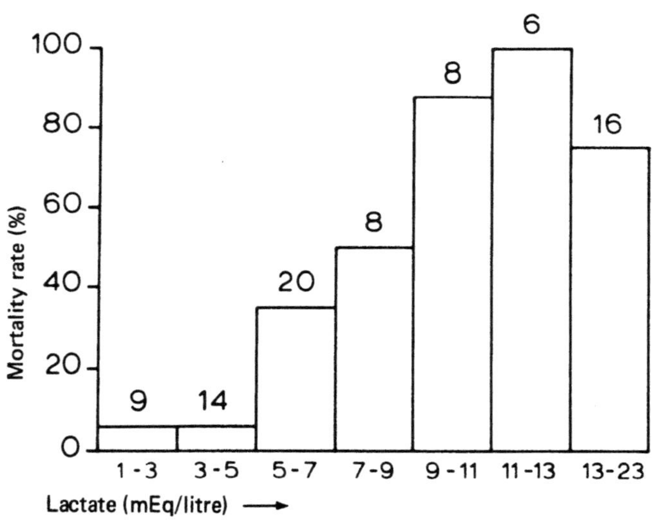

Fig. 3. Mortality rate in shock as a function of the maximal blood lactate level (n = 81)

It is usually accepted that blood lactate accumulation in shock is directly related to tissue production. This has been confirmed by many human and experimental studies. However, it must be realised that the blood lactate level does not only depend on the rate of lactate production, but also on the rate of lactate uptake.

Rowell et al. (6) calculated a rate of lactate clearance as high as 200 mmol/litre during physical exercise in human subjects. Several experiments demonstrated that lactate removal was performed principally by the liver (7-9). Ballinger et al. (10) in experimental studies in dogs, gave evidence of an effective removal of lactate by the liver in the early phase of haemorrhagic shock. Furthermore, in a previous study, we were able to show that hepatic ischaemia could prevent lactate removal in alkalotic dogs (11).

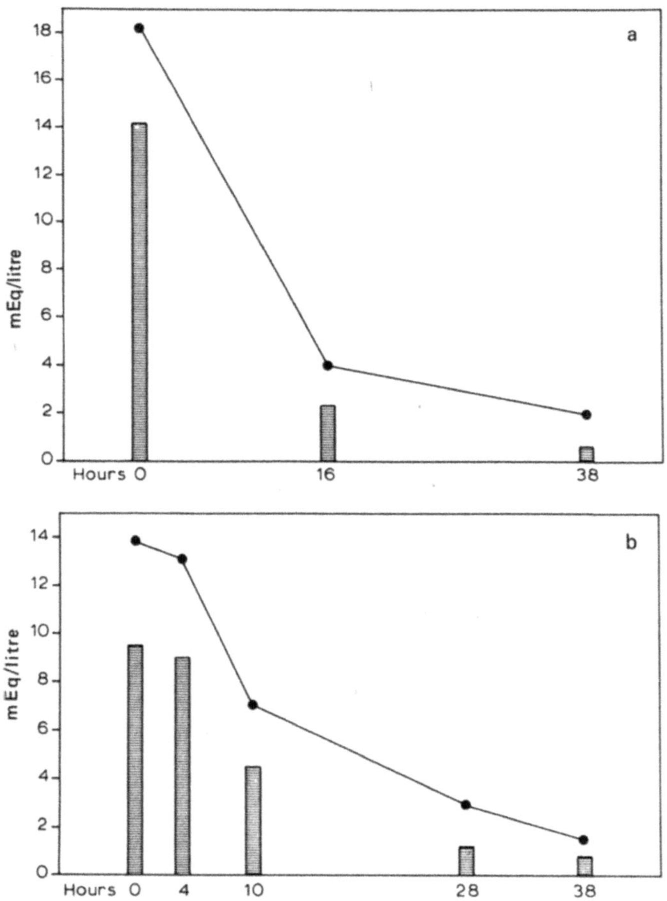

Fig. 4. a and b Course of lactate concentration (•) and excess lactate level (▨) in two patients with liver cirrhosis, both of whom survived haemorrhagic shock

From these data, arterial lactate values observed during shock could be expected to be related not only to peripheral tissue production but also to the liver capacity for removal of these metabolites from the circulating blood. If this is correct, one might expect that hyperlactataemia must be more pronounced when tissue anoxia develops in patients with impaired liver function. The evolution of lactate concentration in two patients with liver cirrhosis is reported in Fig. 4. These two patients had been admitted in haemorrhagic shock following oesophageal varicose rupture. In both patients, blood lactate reached extremely high values, predicting a fatal outcome according to the classic view. However, following haemostasis and volume replacement, blood lactate decreased rapidly and the two patients recovered uneventfully.

Table 1 shows the mean lactate values obtained in hypovolaemic shock, after separating patients with and without liver cirrhosis. The mean lactate value is 7.1 mEq/litre in the patients without liver damage and 13.1 mEq/litre in those with cirrhosis. The difference is significant. The same is true for excess lactate.

The next question is that of the prognostic value of blood lactate in cirrhotic patients in shock. The mean values of the maximal individual lactate level recorded in patients who survived haemorrhagic shock are shown in Table 2. Of 14 patients, nine were cirrhotic and five had no pre-existing liver disease. In patients with shock and cirrhosis, the mean lactate level is very high, while in the other group it is moderately elevated. For excess lactate, the mean value is 7.1 mEq/litre in the cirrhotic patients and 1.75 mEq/litre in the other group. This difference is statistically significant.

<u>Table 1.</u> Mean lactic value in patients in hypovolaemic shock with and without cirrhosis

	Without cirrhosis *n* = 10	Cirrhosis *n* = 18	
Lactate (mEq/litre)	7.1 (1.60)[a]	13.1 (1.24)[a]	$P < 0.05$
XL (mEq/litre)	4.5 (1.18)[a]	9.4 (1.05)[a]	$P < 0.025$

[a] ε

Table 2. Means of individual maximum values for lactate levels in patients surviving haemorrhagic shock

	Without cirrhosis $n = 5$	Cirrhosis $n = 9$	
Lactate (mEq/litre)	3.4 (0.54)[a]	10.5 (1.54)[a]	$P < 0.02$
XL (mEq/litre)	1.75 (0.66)[a]	7.1 (1.41)[a]	$P < 0.05$

[a] ε

Discussion

These results suggest two conclusions: First, high lactate levels have no prognostic value in patients with previous liver disease. In these conditions, because of the decreased lactate liver uptake, a reversible oxygen deficit may lead to severe hyperlactataemia. Second, in patients without liver disease who survive shock, the lactate level rises very moderately.

Now let us consider the opposite situation, that of the patients who died while their maximal blood lactate level remained low (Fig. 3). All these patients had been admitted for septic shock, suggesting a different significance of blood lactate in this condition. Rosenberg und Rush (5) have already reported that blood lactate levels in dogs were lower in endotoxic shock than in hypovolaemic shock, although the estimated cumulative oxygen debt was similar in both conditions. These authors suggested the possible role of a tissue metabolism inhibition secondary to the action of bacterial endotoxin. There might be another explanation for the discrepancy between oxygen deficit and lactate concentration. It is now generally admitted that septic shock provokes particular haemodynamic changes. Numerous experimental and clinical studies have shown that cardiac output during septic shock is frequently normal, or even increased (12-15). This hyperdynamic circulatory state goes on with a particular distribution of cardiac output. Wyler et al. (16) demonstrated in the unanaesthetised monkey that hepatic blood flow was regularly increased in the early phase of septic shock.

From a metabolic point of view, it might be that a hyperdynamic circulatory state with relatively increased liver perfusion could contribute to the maintenance of an adequate oxygen supply to hepatic cells, thus giving proper conditions for an effective removal of lactate. But if shock is prolonged, worsening of the cardiocirculatory situation would lead to progressive hepatic ischaemia, with consequent impairment of liver function.

Thus, if such mechanisms were present, in septic shock the arterial blood lactate level would be expected to depend clearly on the presence or absence of anoxic dysfunction of the liver. To test this hypothesis we divided our patients with septic shock into two groups (Table 3). In the first one, liver function tests never revealed any signs of cytolysis. The absence of hepatic lesions was confirmed at autopsy in all the patients who died. In this group the mean maximal blood lactate level was 3.75 mEq/litre. In the second group, in which all patients showed metabolic and histological signs of extensive hepatic cytolysis, the lactate concentration was increased to 11 mEq/litre. This difference is highly significant. The same is true for excess lactate.

In conclusion, we propose that the blood lactate level in shock is determined by the balance between tissue production and the overall capacity to clear it. The liver appears to be the most effective organ in removal of lactate. Consequently, the lactate level is not only related to cellular anoxia and oxygen deficit, but also to the adequacy of liver function. Liver failure may be the direct consequence of prolonged ischaemia or it may result from previous hepatic damage, as in cirrhotic patients. In the case of earlier hepatic damage, severe

Table 3. Mean lactate values in patients admitted in septic shock, with or without liver damage

	Without liver damage n = 13	With liver damage n = 15	
Lactate (mEq/litre)	3.75 (0.41)[a]	11.0 (1.47)[a]	$P < 0.005$
XL (mEq/litre)	1.5 (0.21)[a]	7.5 (1.09)[a]	$P < 0.001$

[a] ε

hyperlactataemia may develop in the absence of irreversible tissue anoxia. Conversely, when liver perfusion is preserved, as probably occurs in certain cases of septic shock, hepatic removal of lactate is thought to be sufficiently effective to maintain surprisingly low blood lactate levels despite severe tissue anoxia.

Summary

Eighty-four patients with shock syndromes of varied aetiology were investigated. Arterial blood was repeatedly collected for measurements of pH, PCO_2, base excess, lactate, and pyruvate.

Upon admission, metabolic acidosis with various degrees of compensation was the most common disturbance observed (62% of cases). Respiratory alkalosis, with or without a superimposed metabolic component, was noted in the other cases. A significant correlation was obtained between lactate level and base deficit (r = 0.65); $P < 0.001$), showing that acidosis early in shock is largely or entirely due to lactate accumulation.

A clear relationship was noted between the maximal blood lactate concentration and the mortality rate. However, it became apparent that lactataemia is of less prognostic value in cases with pure septic shock and in patients with liver cirrhosis. In the former, a low lactate level (< 3.5 mEq/litre) was accompanied by a high mortality rate; in cirrhosis, exceedingly high lactate levels (> 11 mEq/litre) were observed in patients who survived.

Experimental and clinical arguments suggest that lactataemia in shock is a composite index, which depends not only on the anoxic tissue production but also on the hepatic capacity to clear lactate. In cirrhosis, reduced lactate metabolism by the liver would explain the high lactate levels in reversible shock. Conversely, increased liver perfusion, as observed in septic shock, prevents severe hyperlactataemia, even in cases with a fatal outcome.

References

1. COURNAND, A., RILEY, R.L., BRADLEY, S.E., BREED, E.S., NOBEL, R.O., LAUSON, H.D., GREGERSEN, M.I., RICHARDS, D.W.: Studies of circulation in clinical shock. Surgery 13, 964 (1943)
2. PERETZ, D.I., McGREGOR, M., DOSSETOR, J.B.: Lactic acidosis: a clinically significant aspect of shock. Can. Med. Assoc. J. 90, 673 (1964)
3. BRODER, G., WEIL, M.H.: Excess lactate: an index of reversibility of shock in human patients. Science 143, 1457 (1964)
4. HUCKABEE, W.E.: Relationship of pyruvate and lactate during anaerobic metabolism. II. Exercise and formation of O_2 debt. J. Clin. Invest. 37, 255 (1958)
5. ROSENBERG, J.C., RUSH, B.J.: Blood lactic acid levels in irreversible hemorrhagic and lethal endotoxin shock. Surg. Gynecol. Obstet. 126, 1247 (1968)
6. ROWELL, L.B., KRANING, K.K., EVANS, T.O., KENNEDY, J.W., BLACKMAN, J.R., KUSUMI, F.: Splanchnic removal of lactate and pyruvate during prolonged exercise in man. J. Appl. Physiol. 21, 1773 (1966)
7. SCHIMASSEK, H.: Der Einfluß der Leber auf den extrazellulären Redox-Quotienten Lactat/Pyruvat. Versuche an der isoliert perfundierten Rattenleber. Biochem. Z. 336, 468 (1963)
8. VANG, J.O., DRAPANAS, T.: Metabolism of lactic acid and keto acids by the isolated perfused calf liver: Evaluation of perfusion via portal vein and hepatic artery. Ann. Surg. 163, 545 (1966)
9. BERRY, M.N., SCHEUER, J.: Splanchnic lactic acid metabolism in hyperventilation, metabolic alkalosis and shock. Metabolism 16, 537 (1967)
10. BALLINGER, W.F., VOLLENWEIDER, H., MONTGOMERY, E.H.: The response of the canine liver to anaerobic metabolism induced by hemorrhagic shock. Surg. Gynecol. Obstet. 112, 19-26 (1961)
11. PERRET, C., POLI, S., ENRICO, J.F.: Lactic acidosis and liver damage. Helv. Med. Acta 35, 377 (1969/70)
12. MACLEAN, L.D., MULLIGAN, W.G., McLEAN, A.P.H., DUFF, J.H.: Patterns of septic shock in man. A detailed study of 56 patients. Ann. Surg. 166, 543 (1967)
13. ROSOFF, L., WEIL, M.H., BRADLEY, E.C., BERNE, C.J.: Hemodynamic and metabolic changes associated with bacterial peritonitis. Am. J. Surg. 114, 180 (1967)
14. COHN, J.D., GREENSPAN, M., GOLDSTEIN, C.R., GUDWIN, A.L., SIEGEL, J.H., DEL GUERCIO, R.M.: Arteriovenous shunting in high cardiac output shock syndromes. Surg. Gynecol. Obstet. 127, 282 (1968)
15. ENRICO, J.F., HADORN, R., POLI, S., PERRET, C.: Intérêt de l'exploration hémodynamique au cours du choc septique. In: Journées de réanimation de l'hôpital Claude-Bernard, (pages 146) Paris: Librairie Arnette 1971
16. WYLER, F., FORSYTH, R.P., NIES, A.S., NEUTZE, J.M., MELMON, K.L.: Endotoxin-induced regional circulatory changes in the unanesthetized monkey. Circ. Res. 24, 777 (1969)

Chapter 15

Peripheral Lactic Acid Production in Heart Disease

R. Messin

The present study is limited to lactate production in connection with exercise.

Significance of Exercise-Induced Lactacidaemia in Normal Subjects

Aerobic cellular metabolism depends on the efficiency of the oxygen-transporting system, the cardiorespiratory apparatus playing a prominent role.

The first and immediately available energising source consists in adenosine triphosphate (ATP), whose hydrolysis into adenosine diphosphate (ADP) and phosphate liberates the chemical energy needed by the muscular contraction mechanism, among others. Resynthesis of ATP then takes place immediately, thanks to the energy released by the hydrolysis of creatine phosphate (CP) into creatine and phosphate. The ATP-CP or phosphagen system depends quantitatively on muscular masses; it represents a high energising power released by anaerobic alactic catabolism and having about three times the aerobic power (1, 2).

When phosphagen has delivered its energy, muscular glycogen catabolism becomes responsible for ATP resynthesis, with simultaneous production of pyruvate in the presence of nicotinamide adenine dinucleotide (NAD), a phosphonucleotide accepting H^+ ions.

If oxygen is available, glycogenolysis progresses aerobically through the Krebs cycle, until pyruvate is catabolised into water and CO_2, the energy so released being used for the resynthesis of new molecules of ATP.

The role of oxygen is threefold:

1) Regeneration of NAD by chemical reduction, enabling resynthesis of ATP to persist through aerobic glycogenolysis
2) Maintenance of the Krebs cycle reactions, ensuring ATP resynthesis from ADP, phosphate, and pyruvate, or even, during exercise, from fatty acids and lactate
3) Provision for the resynthesis of glycogen from lactate both during exercise and during the recovery period, when 15% of the lactic acid production would be classically oxidised, accounting partly for the oxygen debt phenomenon and making neoglycogenesis possible from the remaining 85% (3, 4). Studies in animals with labelled lactate seem to indicate, however, that the fate of lactic acid after exercise is primarily oxidative (5).

When the oxygen supply to the tissues is not sufficient to assure all the glycogenolysis required by cellular needs in ATP, the latter is resynthesised by lactic glycogenolysis; pyruvate produced by glycogen catabolism replaces the oxygen as a reducing agent regenerating NAD and is consequently converted into lactate.

Anaerobic glycogenolysis depends quantitatively on muscular glycogen stores, which show large differences according to subjects and circumstances; it represents and energising power about one-and-a-half to two times the aerobic power (1, 2). Peripheral oxygen delivery may be inadequate in several physiological circumstances:

1) At the beginning of submaximal exercise, oxygen uptake per time unit at the pulmonary capillary bed level does not instantaneously reach its steady-state value. The kinetics of that adaptation depends (a) on the amount and speed of utilisation of blood and myoglobin stores in oxygen; (b) on the relative speed of aerobic and anaerobic glycogenolysis leading to ATP resynthesis; and (c) on the adaptation speed of cardiac output (6). After an initial increase, blood lactate concentration falls progressively, provided that exercise is light enough, (a) through diffusion to extracellular spaces; and (b) through oxidation in the Krebs cycle actually during exercise, as mentioned previously (Fig. 1).

2) During the steady-state phase of submaximal exercise reaching at least 50%-70% of the aerobic power, depending on the degree to

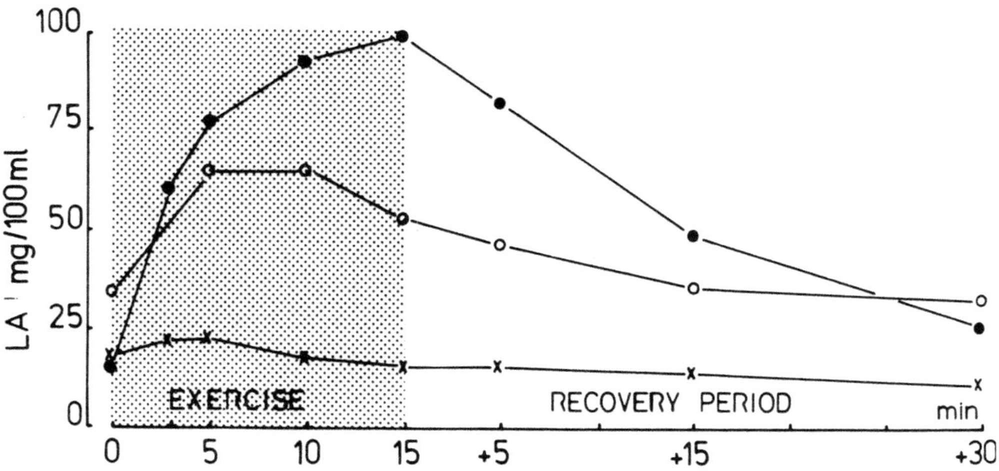

Fig. 1. Lactate concentration in plasma (LA) during and after exercise on the bicycle ergometer in the sitting position; blood was taken from the brachial artery. Lower, middle and upper curves correspond respectively to light, moderate and heavy exercise

which muscle masses are stressed as a whole, lactic glycogenolysis takes place continuously so that blood lactate concentration increases faster and to a greater degree as the work level becomes higher (Fig. 1).

The transition from holoaerobic to meroaerobic work is accompanied by a more rapid rise of ventilation and respiratory exchange ratio due to a larger release of CO_2 from bicarbonate by the metabolic acidosis (7, 8).

3) When exercise intensity reaches or goes beyond the aerobic power, the oxygen-transporting power of the cardiorespiratory system being exceeded, a huge production of lactate takes place with a blood concentration of 100-150 mg/100 ml (or 11-17 mmol/litre) (Fig. 1). On the other hand, the quantitative role of lactic glycogenolysis is more marked than that of aerobic glycogenolysis, as maximal exercise is more quickly reached (9).

4) At altitude, the maximal oxygen-transporting power of the cardiorespiratory system is reduced owing to the decreased alveolar

oxygen partial pressure, and thus the arterial blood desaturation. This results in a lactate production that is significantly increased over that at sea level for the same exercise power (10).

Lactacidaemia and Relative Exercise Intensity in Normal Subjects

The amount of lactic glycogenolysis during physical exercise depends, in fact, on the intensity that submaximal effort represents relative to the maximal individual power, as measured by the maximal oxygen consumption or aerobic power. When blood lactate concentrations corresponding to increasing work levels are expressed in this way, scattering of values is minimal, regardless of individual physical fitness. In the particular case of an artificially decreased aerobic power, e.g., through altitude exposure, blood lactate concentrations far higher than are observed at sea level with the same submaximal work loads, as mentioned previously, become very similar to sea-level values when identical maximal oxygen consumption per centages are considered whose absolute values differ by as much as 30% (9).

Factors Conditioning Exercise-Induced Lactacidaemia in Chronic Heart Disease

We performed a progressive exercise test on the bicycle ergometer in ten controls and in well-compensated patients in sinus rhythm with mitral stenosis (27 cases), mitral disease (6 cases), aortic disease (10 cases), or myocardial infarction with or without angina (respectively, 22 and 47 cases). Respiratory parameters were recorded by a closed-circuit method with a large Knipping's spirograph; cardiac output was measured by the dye-dilution method of Stewart-Hamilton, with indocyanine-green as an indicator, and plasma lactate concentrations were determined enzymatically in blood taken from the brachial artery (11). The workload was increased by 25 W every 5 min until subjective symptoms of intolerance appeared: dyspnoea, angina pectoris, or intense muscular fatigue, or until a criterion for stopping the test was fulfilled: marked electrocardiographic signs of ischaemia, severe rhythm disturbances, or acceleration of heart rate to about 85% of the maximal heart rate estimated from age, this being about 220 minus age. The highest oxygen consumption reached in this test, the so-called $\dot{V}_{O_2}$ SL-CL or oxygen consumption symptom (12), or criterion limited, (symptom limited-criterion limited) measures the functional capacity of the subject. In controls, it corresponds to about 80% of the aerobic power (13).

$\dot{V}O_2$ SL-CL (expressed per kilogram body weight) was reduced to varying extents in the several groups of cardiac patients, owing to their limited maximal oxygen-transporting power; it was significantly lower in mitral valvular disease and in cases of myocardial infarction. Blood lactate concentrations corresponding to the $\dot{V}O_2$ SL-CL were not significantly different from those of controls on average, although the scatter of individual values was rather wide in all groups. The lowest mean value was observed in the group of patients who had myocardial infarction with angina, where exercise had to be interrupted owing to myocardial ischaemia long before the oxygen-transporting system was maximally stressed (Fig. 2).

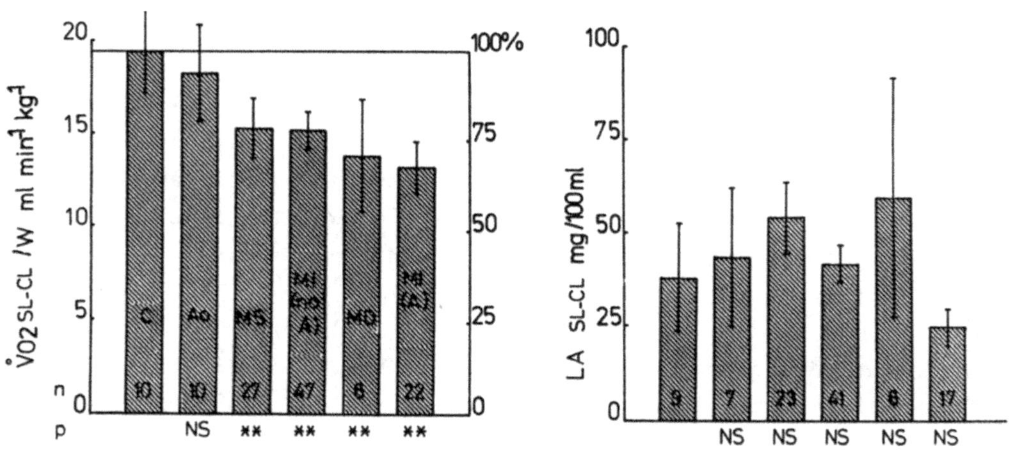

Fig. 2. Oxygen consumption (per kilogram body weight) limited by a symptom or by a criterion for stopping the exercise test ($\dot{V}O_2$ SL-CL) and corresponding lactate concentration in plasma (LA SL-CL), in the different groups investigated. Exercise was performed on the bicycle ergometer in the sitting position; blood was taken from the brachial artery.
Mean values ±2 SE are expressed in ml/(min x kg) (left = hand scale) and as percentages of that obtained in controls (right-hand scale). Comparisons with the control group were made with the t-test of Student-Fisher; the nonparametric Wilcoxon-White test was used for myocardial infarction with angina, owing to the non-gaussian distribution of lactate concentrations in that group.
C, controls; Ao, aortic disease; MS, mitral stenosis; MI (no A), myocardial infarction without angina; MD, mitral disease; MI (A), myocardial infarction with angina. NS, not significant, **, significant at the 1% probability level

The commonest reason for a reduced functional capacity was a decrease in the oxygen-transporting power of the cardiorespiratory system (except in the group of patients with angina) related to a functional alteration of some of its components, which was specific for each heart disease. This is shown by a comparative analysis of the respiratory, haemodynamic, and metabolic parameters recorded at several work levels (13). The lag in oxygen uptake adaptation demonstrated at the beginning of exercise in mitral and aortic diseases (14) probably plays an additional role.

In mitral valvular disease, there was a simultaneous alteration of respiratory function (decrease in ventilatory reserve and pulmonary diffusing capacity, heterogeneity of ventilation:perfusion ratios) and of circulatory function (decrease of stroke volume, cardiac output, and some regional blood flows). It is worth noticing, however, that the relative reduction of splanchnic blood flow in mitral stenosis was the same as in controls when related to the percentage of maximal oxygen consumption used (15). In this group of patients, the lag in cardiac output adaptation at the beginning of exercise contributes to the development of hyperlactacidaemia (16).

In aortic valvular disease, functional alteration of the oxygen-transporting system was much less marked, cardiac output being well maintained with only a slight reduction of stroke volume, perhaps thanks to the left ventricular hypertrophy induced by increased internal mechanical work during the ejection phase. The functional capacity was also little altered, as is shown by the mean $\dot{V}O_2$ SL-CL obtained in this group (Fig. 2).

In myocardial infarction without residual angina, the reduction of the oxygen-transporting power is due to a decrease in maximal cardiac output resulting from a slight decrease of stroke volume, perhaps related to altered myocardial contractility, and to a lower maximal heart rate due to the fact that our patients were an average of 10 years older than the controls.

In myocardial infarction with angina, the lack of adaptation of coronary blood flow was the main reason exercise had to be stopped before other significant haemodynamic disturbances took place.

As a consequence of the reduced oxygen-transporting power, lactate production at submaximal exercise levels was higher in valvular and ischaemic heart disease without angina than in controls (Fig. 3).

Lactacidaemia and Relative Exercise Intensity in Chronic Heart Disease

We aimed to verify whether the relative exercise stress laid on the circulation stimulates lactic glycogenolysis in heart disease in the same way as in normal subjects. Therefore, we related each submaximal blood lactate concentration to the corresponding fraction of in-

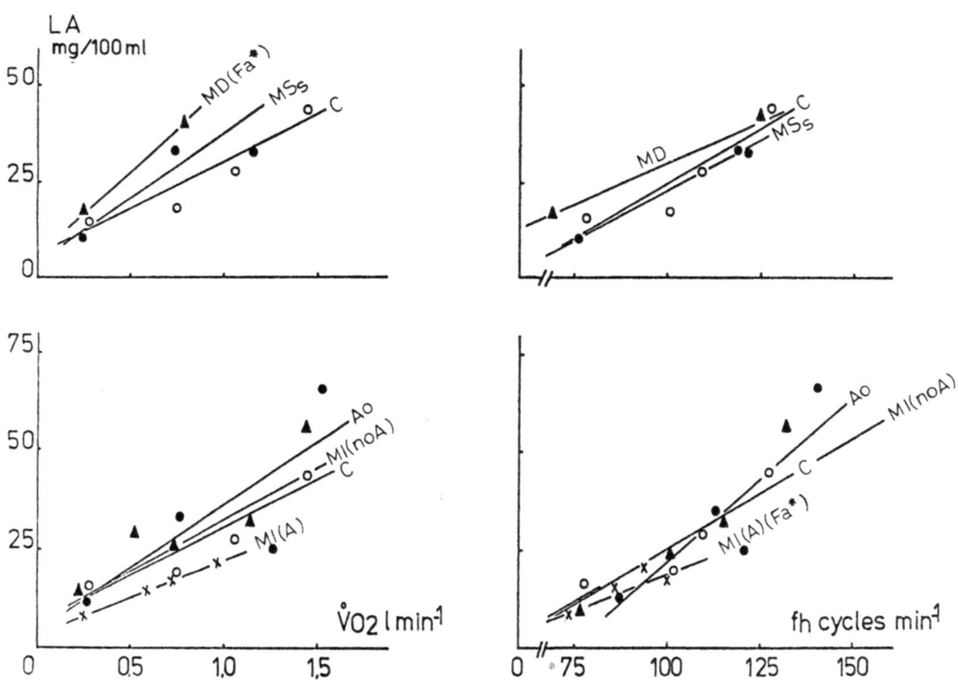

Fig. 3. Regression lines of lactate concentration in plasma (LA) on oxygen consumption (V_{O_2}) or on heart rate (f_h) at rest and at submaximal work loads in the different groups investigated. Exercise was performed on the bicycle ergometer in the sitting position; blood was taken from the brachial artery. Dots and crosses indicate class means at rest, at 40 and 70 W and (for some groups) at 25 or 100 W. Comparisons with the control group are made by analysis of variance and covariance. C, controls; MS_s, mitral stenosis in sinus rhythm; MD, mitral disease; Ao, aortic disease; MI (no A), myocardial infarction without angina; MI (A), myocardial infarction with angina; number of measurements: C, 56; MS_s, 27; MD, 8; Ao, 17; MI (no A), 68; MI (A), 28. $F_a^{\star}$ = F-test showing a difference significant at the 5% probability level between adjusted means

dividual maximal exercise power, used, which can be expressed by the heart rate level reached. In this way, regression lines for blood lactate concentrations are close to those observed in controls, except in for the patients with myocardial infarction with angina; in this group, heart rates are particularly low and the oxygen consumption reached when myocardial ischaemia appears does not correspond at all, of course, to the maximal oxygen-transporting power of the cardiorespiratory system (Fig. 3). The role of relative exercise intensity as a determinant of the lactacidaemia level is further illustrated by the significantly smaller scattering of mean blood lactate concentrations obtained in controls as well as in cardiac and pulmonary patients, when related to relative instead of absolute work levels (Fig. 4).

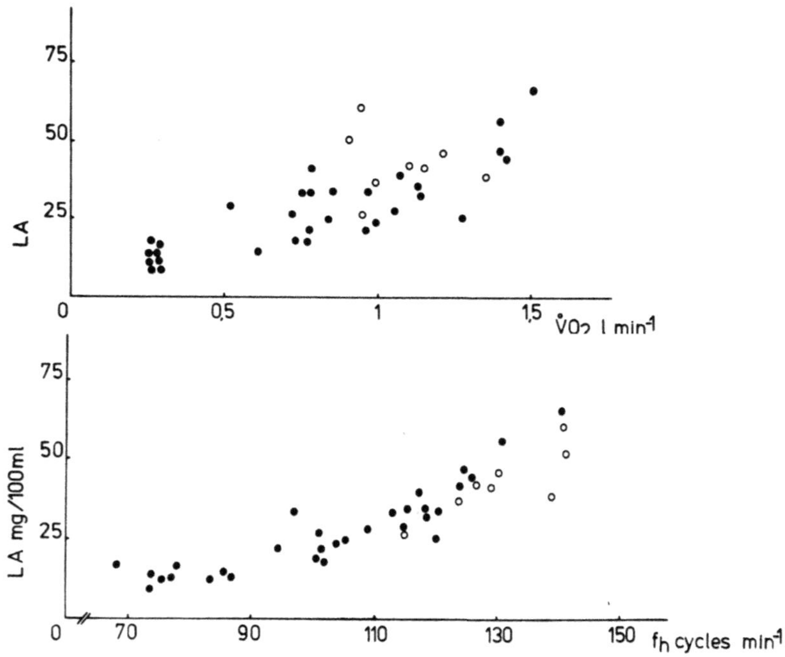

Fig. 4. Relationship between mean lactate concentration in plasma (LA) and oxygen consumption ($\dot{V}_{O_2}$) or heart rate (f_h) at rest and during exercise in controls and in patients with chronic heart or lung disease. Exercise was performed on the bicycle ergometer in the sitting position; blood was taken from the brachial artery.
Filled circles correspond to 31 means of 2-32 measurements performed at rest and at the fifth minute of submaximal exercise levels in the following groups: controls, mitral stenosis in sinus rhythm, mitral disease, aortic disease, left-to-right shunts, myocardial infarction without and with angina, anthracosilicosis, emphysema and/or chronic bronchitis, chronic diffuse interstitial fibrosis.
Open circles correspond to eight means of 2-40 measurements performed when oxygen consumption limited by a symptom or by a criterion for stopping the test was reached; same groups as for submaximal exercise levels, except left-to-right shunt and chronic diffuse intersitial fibrosis

Conclusion

The same relationship between exercise-induced lactacidaemia and the relative percentage of the individual maximal exercise power is observed in controls and valvular or ischaemic heart disease provided that impairment of the oxygen-transporting system includes the systemic circulation as a limiting factor of exercise performance.

Summary

Elevation of the blood lactate concentration during exercise is conditioned by anaerobic glycogenolysis taking part in the resynthesis of high-energy phosphates or phosphagen.

During the period of adaptation to submaximal exercise, lactic acid production increases with the exercise level; it depends on the delay for cardiac output adjustment and for intervention from the aerobic glycogenolytic process. Thereafter, the blood lactic acid concentration decreases progressively, owing to diffusion towards the extracellular spaces and to oxidation in the Krebs cycle with resynthesis of ATP or glycogen. With increasing workloads, a continuous large production of lactate takes place from 50%-70% of the maximal oxygen consumption; consequently, lactacidaemia rises rapidly, reaching 100-150 mg/100 ml (or 11-17 mmol/litres) when the maximal oxygen-transporting capacity of the cardiorespiratory system is exceeded. The role of anaerobic glycogenolysis becomes more important compared with that of aerobic glycogenolysis as maximal exercise becomes more rapidly exhaustive. The energising power of lactic glycogenolysis is on average one-and-a-half to two times the aerobic power.

Controls and patients in sinus rhythm with mitral stenosis, mitral disease, aortic disease and myocardial infarction with or without angina were subjected to an exercise test on the bicycle ergometer with progressively increasing workloads while respiratory, haemodynamic and blood parameters were registered. Functional capacity is defined as the oxygen consumption limited by a symptom or by a criterion for stopping the test ($\dot{V}_{O_2}$ SL-CL); in normal subjects the latter corresponds to about 80% of the aerobic power. Whether the factor interfering with oxygen transport to the working organs is lack of adaptation of cardiac output (as in aortic insufficiency or in myocardial infarction without angina) or alteration of the circulatory and respiratory functions (as in mitral valvular disease), the increase of lactate in arterial blood depends, just as in normal subjects, on the fraction of maximal exercise power used; that fraction is proportional to the heart rate level reached. Consequently, for the same submaximal oxygen consumption, lactacidaemia becomes more pronounced as the cardiorespiratory possibilities of oxygen transport are reduced and cardiac output adaptation is slower, as in mitral stenosis. On the other hand, when $\dot{V}O_2$ SL-CL is reached, the blood lactate concentration in the above-mentioned groups with heart disease is not significantly different from that in the controls. In the group of patients with myocardial infarction with angina, the oxygen comsumption corresponding to myocardial ischaemia does not reflect the oxygen-transporting power of the cardiorespiratory system.

References

1. MARGARIA, R., CERRETELLI, P., MANGILI, F.: Balance and kinetics of anaerobic energy release during strenuous exercise in man. J. Appl. Physiol. 19, 623-628 (1964)
2. FLANDROIS, R., LACOUR, J.R.: Evolution des paramètres métaboliques et respiratoires au cours de l'exercise musculaire chez l'homme. Bull. Eur. Physiopathol. Respir. 13, 329-341 (1977)
3. KREBS, H.: Discussion. In: Oxygen in the animal organism. Dickens, F., Neil, E., Widdas, W.F. (eds.), pp. 304-305. New York: Pergamon Press 1964
4. HERMANSEN, L.: Lactate metabolism during and after exercise. In: Symposium sur le métabolisme anaérobique de l'acide lactique. Organisé sous le patronage de la Société Européenne de Cardiologie, Anzère (Valais), 19-22 janvier 1978.
5. BROOKS, G.A., BRAUNER, K.E., CASSENS, R.G.: Glycogen synthesis and metabolism of lactic acid after exercise. Am. J. Physiol. 224, 1162-1166 (1973)
6. CERRETELLI, P., di PRAMPERO, P.E.: Kinetics of respiratory gas exchange and cardiac output at the onset of exercise. Scand. J. Respir. Dis. [Suppl.] 77, 35a-35g (1971)
7. CHERCHI, A.: A synthetic triangular test. In: Ergometry in cardiology. Denolin, H., König, K., Messin, R., Degré, S. (eds.), pp. 65-86. Mannheim: Boehringer 1968
8. CLODE, M., CAMPBELL, E.J.M.: The relationship between gas exchange and changes in blood lactate concentrations during exercise. Clin. Sci. 37, 263-272 (1969)
9. ÅSTRAND, P.O., RODHAL, K.: Textbook of work physiology. New York: McGraw Hill 1970
10. STENBERG, J., EKBLOM, B., MESSIN, R.: Hemodynamic response to work at simulated altitude 4000 m. J. Appl. Physiol. 21, 1589-1594 (1966)
11. DE COSTER, A., MESSIN, R., FRANCKSON, J.R.M.: Répartition de l'acide lactique entre plasma et globules rouges chez l'homme. Comparaisons des résultats fournis par différentes méthodes de dosage. Arch. Int. Physiol. Biochim. 73, 251-258 (1965)
12. DETRY, J.R.M., BRUCE, R.A.: Effects of physical training on exertional ST segment depression in coronary heart disease. Circulation 44, 390-396 (1971)

13. MESSIN, R.: Analyse statistique et informatique par miniordinateur de grandeurs cardio-respiratoires et métaboliques enregistrées au repos et à l'effort en vue de définir des facteurs limitatifs de la performance physique dans divers types de cardiopathies et pneumopathies chroniques. To be published.
14. AUCHINCLOSS, J.H., Jr., GILBERT, R., BAULE, G.H.: Unsteady state measurement of oxygen transfer in patients with rheumatic heart disease. Clin. Sci. 39, 21-37 (1970)
15. BLACKMON, J.R., ROWELL, L.D., KENNEDY, J.W., TWISS, R.D., CONN, R.D.: Physiological significance of maximal oxygen intake in "pure" mitral stenosis. Circulation 36, 497-510 (1967)
16. MESSIN, R., DEGRE, S., VANDERMOTEN, P., DEMARET, B., DE COSTER, A., DENOLIN, H.: Effect of exercise on cardiorespiratory function and metabolism in mitral stenosis. G. Ital. Cardiol. 20, 636-643 (1972)

Chapter 16

Blood Lactate Measurement in Peripheral Arterial Disease

N. Krähenbühl

Introduction

An increased lactate concentration in venous return from a working muscle is considered a sign of cellular ischaemia. In peripheral arterial disease, measurement of the blood lactate may be useful in the diagnosis and follow-up of occlusive arterial disease; furthermore, in McArdle's disease, which is characterised by a lack of muscle phosphorylase, blood lactate does not rise after exercise sufficient to cause ischaemia.

Diagnosis of Peripheral Arterial Occlusive Disease

In peripheral arterial occlusive disease, muscles are working under ischaemic conditions, so that the venous lactate concentration may be of diagnostic value. However, maximal work in untrained subjects does in fact provoke an increase of venous lactate in the absence of any arterial occlusive disease. This rise in blood lactate after exercise in normal subjects has been studied by Pernow et al. (1). Muscle biopsies before and after exercise were performed in normal subjects and in patients with intermittent claudication (Fig. 1). There was no difference at rest. After gentle exercise there was a moderate increase of muscle lactate in normal subjects, but maximal exercise produced a very significant increase in lactate. In patients with intermittent claudication, maximal exercise, i.e., a walk resulting in muscular ischaemic pain, provoked a similar increase in the muscular lactate concentration.

Therefore, it is necessary to determine the normal values of blood lactate increase after exercise. Carlson and Pernow (2) have compared arterial and venous lactate concentrations in normal subjects and in patients with peripheral arterial occlusive disease at rest and after

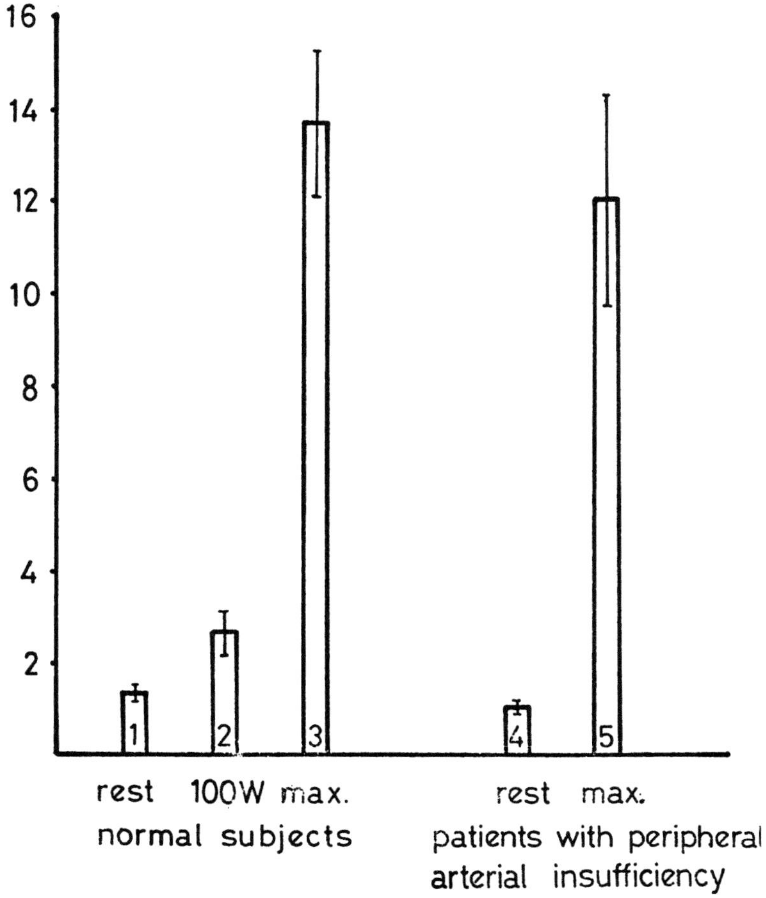

Fig. 1. Concentration of muscle lactate at rest and after exercise. Differences between 1 and 4 and between 3 and 5 are not significant. Difference between 2 and 5 is significant ($P < 0.001$). Pernow et al., 1975

different degrees of work (Fig. 2). Up to 140 beats/min, there was a significant difference between normal subjects and patients with intermittent claudication. This difference is significant both in arterial and in venous blood.

Köhler (3) measured pyruvate and lactate blood concentrations and calculated the "excess lactate" according to Huckabee (4). He found a striking difference between normal subjects and patients with intermittent claudication (Fig. 3).

Thus, measurement of the venous lactate concentration after exercise may be useful in the diagnosis of peripheral arterial occlusive disease. However, in clinical practice, noninvasive haemodynamic

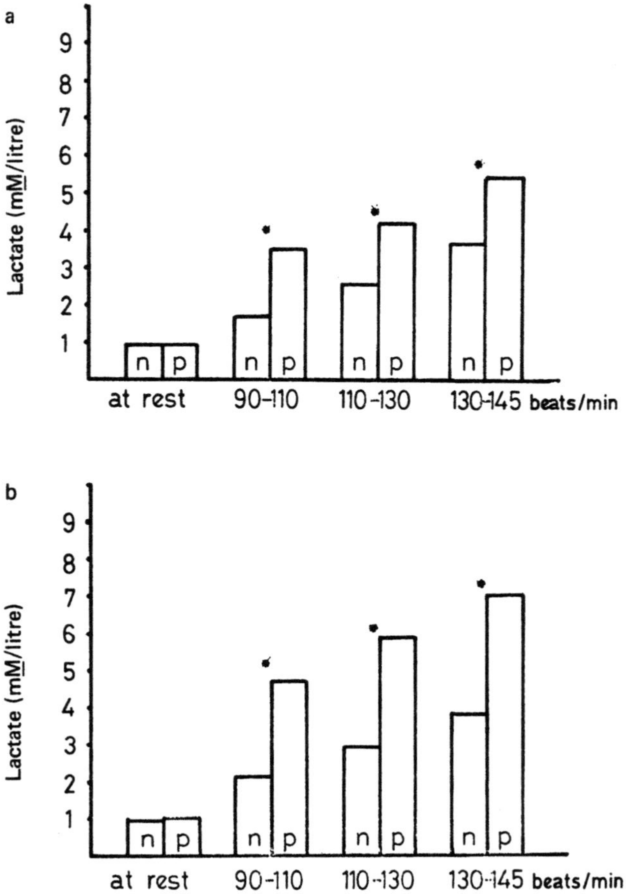

Fig. 2. a and b Lactate levels in femoral artery (a) and femoral vein (b) at rest and after exercise in normal subjects (n) and in patients with peripheral arterial insufficiency (p). ★ denotes a statistically significant difference ($P < 0.001$) between normal subjects and patients. Carlson and Pernow, 1962

measures, especially distal systolic pressure measurement, can be used for the diagnosis of peripheral arterial occlusive disease, so that measurement of blood lactate is seldom necessary.

Follow-Up of Patients with Intermittent Claudication

Arterial surgery constitutes the most effective symptomatic treatment of intermittent claudication. Bypass procedures or arterial clearance usually result in a significant improvement of peripheral arterial systolic pressure, which can be measured by noninvasive techniques (5). The effect of arterial surgery on muscle lactate was studied in 11 patients by Pernow (1). The lactate concentration in muscle at

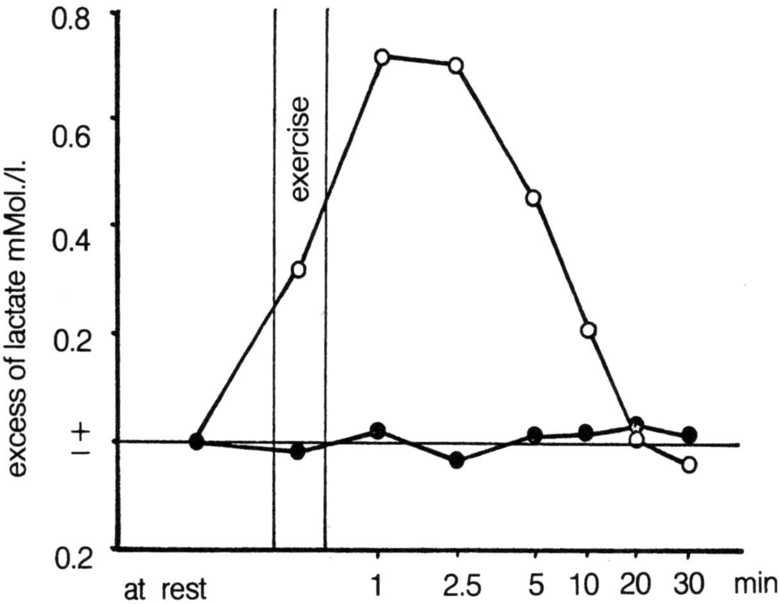

Fig. 3. Excess of lactate in five normal subjects (•) and in 16 patients with peripheral arterial insufficiency (o). Köhler, 1975

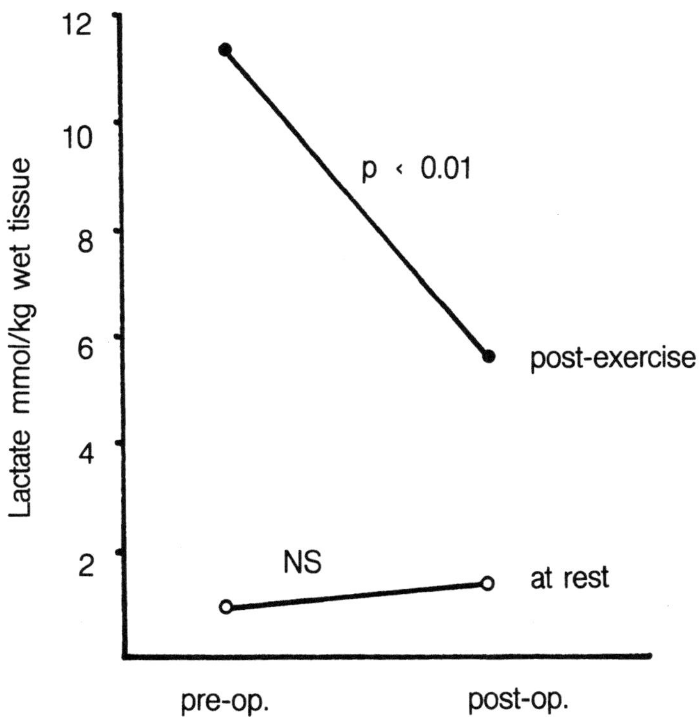

Fig. 4. Effect of exercise on muscle lactate before and after surgery in 11 patients with peripheral occlusive disease. Pernow, 1975

rest did not change after surgery, but the rise after exercise was significantly lower (Fig. 4). Thus, there is a logical relationship between peripheral haemodynamic improvement after surgery and the lactate concentration in muscle.

Conservative treatment may also improve the walking distance in patients with intermittent claudication: after a few weeks of daily walking exercises, walking performance is often much better. Despite a significant improvement in walking distance, there is generally no detectable change in haemodynamic parameters: the arm-to-ankle systolic pressure gradient does not improve. Dahllöf et al. measured the maximal calf blood flow by the xenon clearance technique before and after 3 months of training in a group of patients with intermittent claudication (6). The mean walking tolerance rose from 300 to 600 m, but the maximal calf blood flow did not change. However, the muscle lactate concentration fell significantly after the training period (Fig. 5). This clinical improvement without any detectable haemodynamic change may be explained by a muscular metabolic adaptation to ischaemic work. However, haemodynamic changes that were not detectable by the methods available may also be suspected.

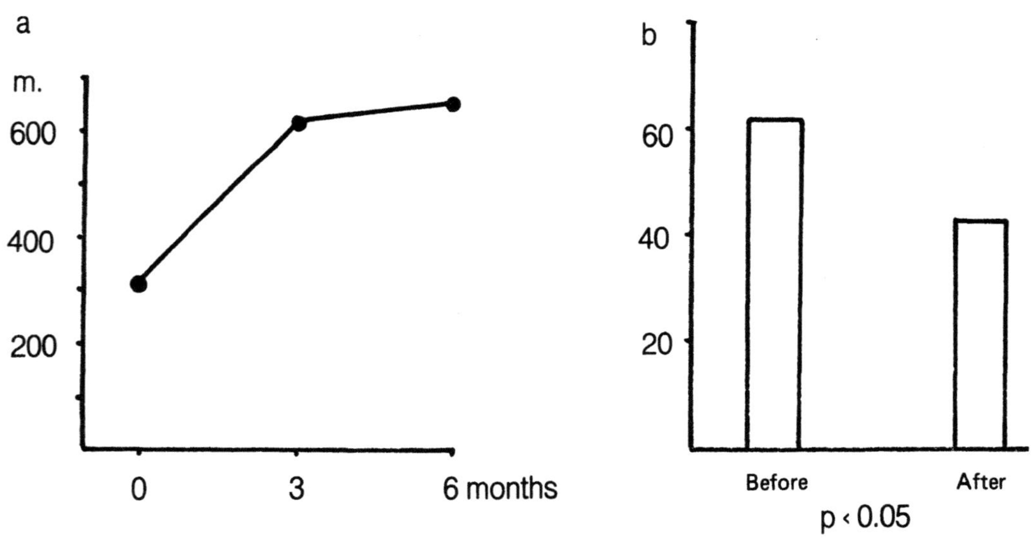

Fig. 5. a and b Effect of training on walking tolerance (a) and muscle lactate (b) in patients with peripheral occlusive disease. Calf muscle blood flow was unchanged. The difference in muscle lactate was statistically significant ($P < 0.05$) Dahllöf et al., 1975

McArdle's Disease

Almost 30 years ago, McArdle described a patient suffering from muscle pain and weakness after any muscular exercise (7). A careful examination did not show any vascular disease, and a defect in muscular metabolism was suspected. Since that time, these patients have been shown to be lacking in muscle phosphorylase activity. Although McArdle's disease is very rare, this diagnosis must be considered when a young patient suffers from intermittent claudication in the absence of any arterial disease. When McArdle's disease is suspected, measurement of the venous lactate concentration after an ischaemic work is a useful diagnostic test: in normal subjects, there is a significant rise in venous lactate; in patients with McArdle's disease, the blood lactate does not increase, and may sometimes fall (Fig. 6).

Conclusion

The blood lactate concentration measured after moderate exercise is significantly higher in patients with intermittent claudication than

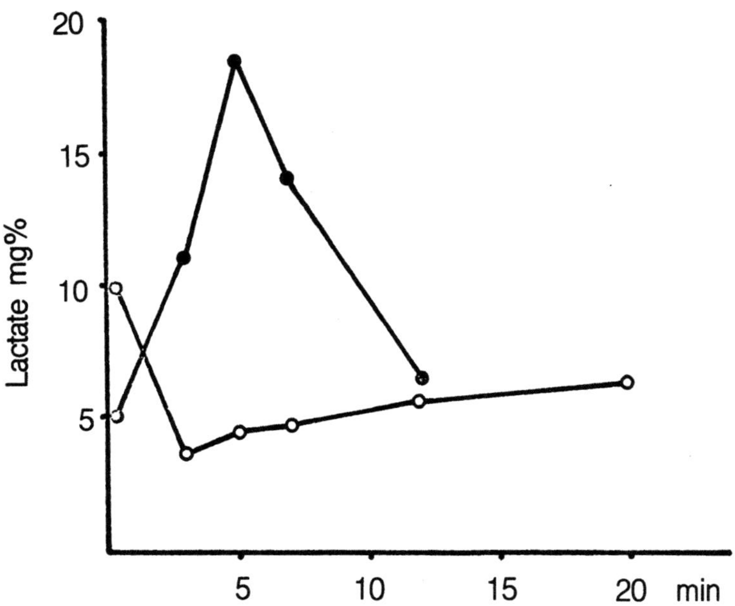

Fig. 6. Absence of blood lactate increase after ischaemic exercise in a patient with McArdle's disease (o). Control values (•) are shown for comparison. McArdle, 1951

in normal subjects. However, this measurement is not generally used in clinical practice, mainly because peripheral arterial occlusive diseases are accurately detected by noninvasive haemodynamic measures, especially distal systolic pressure measurement. Studies of muscular lactate metabolism have provided a possible explanation for the improvement in patients treated by walking exercises: a metabolic adaptation of skeletal muscle to ischaemic work. Finally, when Mc-Ardles disease is suspected, the absence of an increase in venous lactate after an ischaemic exercise is of diagnostic value.

References

1. PERNOW, B., SALTIN, B., WAHREN, J., CRONESTRAND, R., EKESTRÖM, S.: Leg blood flow and muscle metabolism in occlusive arterial disease of the leg before and after reconstructive surgery. Clin. Sci. Mol. Med. 49, 265 (1975)
2. CARLSON, L.A., PERNOW, B.: Studies on the peripheral circulation and metabolism in man. II. Oxygen utilisation and lactate-pyruvate formation in the legs at rest and during exercise in patients with arteriosclosis obliterans. Acta Med. Scand. 171, 311 (1962)
3. KÖHLER, M.: Metabolische und hämodynamische Befunde unter submaximaler und maximaler Belastung. In: Ergometrie und Ergotherapie bei arteriellen Durchblutungsstörungen, Bollinger, A., Gruntzig, A. (eds.), p. 66. Bern: Huber 1975
4. HUCKABEE, W.E.: Relationships of pyruvate and lactate during anaerobic metabolism. I. Effects of infusion of pyruvate or glucose and of hyperventilation. J. Clin. Invest. 37, 244 (1958)
5. CARTER, S.A.: Clinical measurement of systolic pressures in limbs with arterial occlusive disease. JAMA 207, 1869 (1969)
6. DAHLLÖF, A.G., HOLM, J., SCHERSTEN, T., SIVERTSSON, R., CARLSÖÖ, S.: Effect of controlled training with respect to walking tolerance, maximal calf muscle blood flow, gait technique and muscle metabolism in patients with peripheral arterial insufficiency. In: Ergometrie und Ergotherapie bei arteriellen Durchblutungsstörungen. Bollinger, A., Gruntzig, A. (eds.), p. 114. Bern: Huber 1975
7. McARDLE, B.: Myopathy due to a defect in muscle glycogen breakdown. Clin. Sci. 10, 13 (1951)

Chapter 17

Lactic Acid and Diabetes

U. Keller and W. Berger

Introduction

Abnormalities of blood lactate concentrations in diabetes have been described during the past years in an increasing number of reports. A major reason for the gain in interest in lactic acid metabolism has been the recognition of the clinical syndrome of lactic acidosis, which is particularly related to diabetes. In many cases biguanide therapy has to be implicated as the underlying cause. The factors leading to altered blood lactate levels are manifold; the present paper will focus on those that are of specific importance for the production of hyperlactataemia in diabetics.

Lactate Metabolism in Diabetes: Basic Information

It might be perplexing at first to state that diabetes per se does not necessarily lead to abnormal blood lactate concentrations. According to Tranquada et al. (1), lactate levels in well-controlled hospitalised diabetic patients without acute illness were not significantly different from those in controls. In contrast, when the hepatic uptake of lactate was determined in diabetic patients by arteriovenous catheterisation, it was approximately 50% higher than in controls (2). The difference was even more pronounced following ingestion of a protein meal. Since hepatic glucose production was normal in diabetics, a major part of hepatic glucose production in these subjects was derived from gluconeogenic precursors such as lactate. Circulating blood lactate levels did not reveal changes in the hepatic lactate uptake. Therefore, blood lactate levels do not necessarily correlate with production or removal rates of lactate, and more detailed techniques, such as arteriovenous catheterisation or tracer methods are required to estimate the true rate of lactate metabolism. Whereas the former method has the disadvantage of being invasive, the latter involves methodological difficulties that limit its use. As noted by Kreisberg (3), the determination of lactate turnover rates by infusion of lactate ^{14}C involves a considerable amount of glucose-lactate interconversion, which does not represent net uptake or production of lactate by tissues. Results obtained by infusion of glucose ^{14}C and lactate ^{14}C indicated that as much as 20% of lactate turnover was converted to glucose, and 45% of glucose turnover was derived from recirculating lactate. Oxidation of lactate to CO_2 has been determined as 51% of lactate turnover (4).

A further insoluble problem in the use of lactate ^{14}C as a tracer to assess lactate turnover derives from the fact that lactate radioactivity is diluted prior to release by the tissues, due to intracellular exchange of labelled carbon with oxaloacetate, which represents a branch point of gluconeogenesis and oxidation to acetyl-CoA. As demonstrated by Krebs (5), the isotopic flux of ^{14}C through the oxaloacetate pool is not paralleled by true chemical flux; this leads to an underestimation of gluconeogenesis from lactate and to an overestimation of oxidation to CO_2.

Keeping these limitations in mind, lactate turnover has been found to be about 1.3 mmol/min in diabetics (6), which is in the same range as in normal man (16). No further data on the regulation of lactate turnover in diabetes are available.

In patients with diabetes some data indicate a defect in the oxidation of lactate due to diminished pyruvate oxidation. Infusion of pyruvate in diabetics was associated with higher pyruvate concentrations than in control subjects (7), possibly due to decreased activity of pyruvate dehydrogenase during insulin deficiency (8).

On the other hand, the experimental evidence obtained in vitro indicates that insulin inhibits gluconeogenesis from lactate in isolated liver cells (9). These effects of insulin on gluconeogenesis and pyruvate oxidation might exert opposing effects on blood lactate concentrations. In contrast to insulin, glucagon enhances lactate uptake by isolated liver cells by virtue of its potent stimulatory effect on gluconeogenesis (10). In the presence of both hormones a balance of the antagonistic effects of the two was observed (11).

Lactate Concentrations in Diabetic Coma

Observations in patients in diabetic coma indicate that these effects of insulin and glucagon on hepatic gluconeogenesis from lactate are not major determinants of blood lactate concentrations. Diabetic coma is characterised by low circulating insulin and high glucagon levels, and gluconeogenic rates are high, as suggested by increased rates of hepatic uptake of gluconeogenic precursors. However, these patients do not have decreased, but usually increased, blood concentrations of lactate (12). The observation that the L:P ratio is markedly elevated indicates that the hyperlactataemia is largely due to diminished O_2 consumption by the tissues.

When insulin therapy is instituted, lactate levels gradually return to normal after a short initial rise. However, there remains a group of patients who show persistent hyperlactataemia (12). These patients usually have a serious underlying disorder, and the condition is associated with a high mortality (13). The reasons for the elevation of lactate levels in diabetic coma have not been investigated, but dehydration, diminished tissue perfusion, and decreased pyruvate oxidation are important underlying causes.

Lactic Acidosis in Diabetes

More than 90 years ago, two cases of diabetic coma were described where Kussmaul respiration was present but there was no acetone excretion in either urine or breath (14). The question as to the nature of the unknown acid that was produced in excess by these patients

Table 1. Blood lactate and pyruvate concentrations in maturity-onset diabetics after 10 days of placebo, phenformin[a] or metformin[b] therapy (means ± SEM)

	Placebo (*n*=5)	Phenformin (*n*=5)	Placebo (*n*=5)	Metformin (*n*=5)
Lactate (mmol/litre)	0.69 ± 0.05	1.21 ± 0.15[c]	0.69 ± 0.08	1.17 ± 0.16[c]
Pyruvate (mmol/litre)	0.053 ± 0.007	0.085 ± 0.014[d]	0.051 ± 0.007	0.076 ± 0.013[c]
Lactate:Pyruvate ratio	13.7 ± 1.4	14.6 ± 1.1	14.0 ± 1.2	15.8 ± 1.0

[a] 100 mg 2.5 h and 50 mg 1 h before sampling
[b] 1000 mg 2.5 h and 500 mg 1 h before sampling
[c] P vs placebo < 0.025
[d] P vs placebo < 0.05

Table 2. Lactate kinetics[a] in diabetics during treatment with placebo or phenformin[b] (means ± SEM)[c]

	Placebo	Phenformin
Volume of distribution (litres)	21.8 ± 2.6	28.6 ± 5.0
Endogenous production rate (mmol/min)	0.73 ± 0.07	0.54 ± 0.09
Elimination rate constant ($min^{-1} x10^2$)	4.2 ± 0.8	1.8 ± 0.3[d]
Metabolic clearance (ml/min)	1078 ± 121	494 ± 76[e]

[a] Determined by continuous infusion of 0.02 mmol Na-L-lactate/kg/min
[b] 100 mg daily for 6 days, 100 mg 2.5 h and 50 mg 1 h before lactate infusion
[c] A one-compartment model has been assumed
[d] $P <$ vs placebo 0.05
[e] $P <$ vs placebo 0.025

remained unresolved for years. Lexow, in 1959, described three cases of diabetic acidosis without ketonuria (15), and he suggested that the excess acid was lactic acid. Since then more than 20 cases of lactic acidosis in diabetics have been described (13, 16, 17); of particular interest were these cases whose lactic acidosis developed without evidence of hypoxaemia or shock. This form of lactic acidosis has been called "spontaneous" lactic acidosis (17) and classified as type-B lactic acidosis (19). In a recent report, Wise described a series of 38 patients admitted to hospital with diabetes and hyperlactataemia (19). The majority of the cases with lactic acidosis were taking the biguanide phenformin as an oral hypoglycaemic agent. The patients were usually not ketotic, whereas a distinct group of subjects had concomitant ketoacidosis and lactic acidosis.

Although not widely recognised, the association of biguanide therapy and hyperlactataemia has been known for more than 50 years. Staub was the first to observe an increase in blood lactate concentrations in experimental animals following administration of a guanidine compound, synthalin (20). He felt that because of this side effect the drug was too toxic for clinical use, and similar statements are now being made again.

The association between phenformin therapy and lactic acidosis was first recognised in 1963 (21). Lactic acidosis is usually defined as elevation of blood lactate concentrations to over 7 mmol/litre (1). After a delay of several years due to poor recognition of the condition, a large number of cases have been described. So far more than 300 reports of lactic acidosis during phenformin therapy have been published (22, 23), and the true incidence is probably much higher.

In a recent publication we reported a series of ten patients with lactic acidosis during biguanide therapy (24). In all these cases drug levels were clearly elevated and in most instances this could be ascribed to pre-existing impaired renal function. The condition was frequently preceded by a period of nausea and vomiting, which lead to dehydration and to further deterioration of renal function. In spite of alkali therapy with bicarbonate and haemodialysis, 50% of the cases died.

The high incidence of lactic acidosis during biguanide therapy has led to a re-examination of the value of biguanides in the therapy of diabetes; the occurrence of this serious complication and an increased risk of cardiovascular deaths during treatment with phenformin (25) has led to withdrawal of the drug from the market in the United States.

Effect of Biguanide Therapy on Lactate Metabolism in Diabetics

We studied the effect of therapeutic doses of two frequently used biguanides on lactate metabolism in diabetic outpatients. Table 1 shows blood lactate and pyruvate concentrations in five patients with mature-onset diabetes during phenformin or metformin and placebo therapy of 2 weeks duration. Each drug was administered in a random sequence to each patient. Phenformin therapy was associated with a marked rise in blood lactate and pyruvate concentrations to almost twice the values found during placebo administration. The rise in the L:P ratio was slight and statistically insignificant. During metformin therapy a similar elevation of blood lactate and pyruvate to that seen during phenformin medication occurred. Thus, the two drugs had similar effects on resting blood lactate and pyruvate levels.

To examine the effect of phenformin therapy on lactate metabolism more closely, lactate production and clearance rates were determined in mature-onset diabetics by means of primed constant infusions with Na-L-lactate (Table 2; 26). The resulting blood lactate levels and the knowledge of the infusion rate permitted calculation of the rate of endogenous lactate production (27). The slopes of the lactate concentrations were employed to determine the volume of distribution and the excretion rate constant, assuming a one-compartment model. The data indicated that phenformin therapy did not affect lactate production rates. The metabolic clearance rate of lactate was significantly decreased, demonstrating higher rates of utilisation at given blood lactate concentrations. Thus, the results suggested that hyperlactataemia during phenformin therapy results from underutilisation and not from overproduction of lactate.

The mechanism of action of biguanides remains a controversial issue; the present results are consistent with those of in vitro studies demonstrating diminished rates of gluconeogenesis from lactate in the perfused liver (28). However, they appear to contrast with data recorded in normal man, which reflect increased rates of glucose recycling through lactate with tracer methods (29). The finding of a diminished rate of lactate utilisation is in keeping with a previous report demonstrating a relative decrease of lactate oxidation following phenformin administration (6).

Incidence and Mortality of Lactic Acidosis During Treatment with Various Types of Biguanides

The incidence of lactic acidosis varies considerably with different types of biguanides (Table 3). Many more cases of lactic acidosis have been described during phenformin than during metformin therapy.

Table 3. Incidence and prognosis of lactic acidosis associated with various types of biguanides

Country	Biguanide	No, of cases	Incidence per 1000 treatment years[a]	Mortality
France[b]	Metformin	17	0.016	30%
1968-1977	Phenformin	78	0.23	70%
Switzer-land[c]	Metformin	2	0.08	0
1972-1977	Phenformin	6	0.64	83%
	Buformin	34	0.40	41%

[a] Calculated from amount of drug consumed, as communicated by the manufacturers
[b] Includes all published cases
[c] Data obtained from a retrospective survey in all hospitals in the country

This may in part reflect total consumption; however, data obtained from a large number of patients in France indicate that the incidence of lactic acidosis was sixfold higher during phenformin than during metformin therapy, and a similar picture was obtained in a retrospective survey in the city hospitals of Switzerland. Lactic acidosis during metformin therapy occurred only in the presence of severe renal failure (22).

Not only the incidence, but also the mortality rates were higher during phenformin than during metformin therapy. These differences might be due to varying drug metabolism and tissue distribution, since phenformin is partly incorporated into the liver, where it is metabolised, and only 50% is excreted by the kidney (30), whereas metformin is excreted exclusively in the urine.

The diverging pharmacokinetic properties of the two biguanides seem to predispose to the development of lactic acidosis rather than to influence blood lactate concentrations, which are similar to those shown above.

Summary

Blood concentrations of lactate may be normal in diabetic patients, but rates of hepatic uptake and thus gluconeogenesis from lactate have been shown to be increased. Blood lactate levels are usually elevated when control of diabetes is lost. Dehydration, poor tissue perfusion, and diminished pyruvate oxidation caused by insulin deficiency contribute to the development of hyperlactataemia during severely uncontrolled diabetes. Diabetic coma may reveal concomitant ketoacidosis and lactic acidosis.

Severe hyperlactataemia producing lactic acidosis is particularily prevalent in diabetic patients; it may develop without clinical evidence of circulatory failure as spontaneous lactic acidosis. In most cases, earlier biguanide therapy is implicated. Elevated drug levels are usually found in these patients due to impaired renal function.

Clinical observations indicate that the incidence and prognosis of lactic acidosis varies with different types of biguanides, metformin being implicated less frequently than phenformin. Data are presented to demonstrate that phenformin and metformin therapy lead to a significant rise of resting lactate and pyruvate concentrations. Hyperlactataemia appeared to be the result of decreased removal rather than increased production rates of lactic acid. Regardless of the pathogenesis, the development of lactic acidosis in diabetic patients is a serious complication with a high mortality.

References

1. TRANQUADA, R.E., GRANT, W.J., PETERSON, C.R.: Lactic acidosis. Arch. Intern. Med. 117, 192 (1966)
2. WAHREN, J., FELIG, P., HAGENFELDT, L.: Effect of protein ingestion on splanchnic and leg metabolism in normal man and in patients with diabetes mellitus. J. Clin. Invest. 57, 987 (1976)
3. KREISBERG, R.A.: Glucose-lactate inter-relations in man. N. Engl. J. Med. 287, 132 (1972)
4. FREMINET, A., POYART, C.: Lactate-glucose interrelations, glucose recycling and the cori cycle in normal fed rats. Pfuegers Arch 361, 25 (1975)
5. KREBS, H.A., HEMS, R., WEIDEMANN, M.J., SPEAKE, R.N.: The fate of isotopic carbon in kidney cortex synthesizing glucose from lactate. Biochem. J. 101, 242 (1966)
6. SEARLE, G.L., SIPERSTEIN, M.D.: Lactic acidosis associated with phenformin therapy. Diabetes 24, 741 (1975)
7. MOORHOUSE, J.A.: Pyruvate-tolerance tests in healthy and diabetic subjects. Lancet 1, 689 (1964)
8. WIELAND, O.H., PATZELT, C., LÖFFLER, G.: Active and inactive forms of pyruvate dehydrogenase in rat liver. Eur. J. Biochem. 26, 426 (1972)
9. CLAUS, T.H., PILKIS, S.J.: Regulation by insulin of gluconeogenesis in isolated rat hepatocytes. Biochim. Biophys. Acta 421, 246 (1976)
10. EXTON, J.H., PARK, C.R.: Control of gluconeogenesis in liver. J. Biol. Chem. 243, 4189 (1968)
11. PARILLA, R., JIMENEZ, J., AYUSO-PARILLA, S.: Glucagon and insulin control of gluconeogenesis in the perfused isolated rat liver. Effects on cellular metabolite distribution. Eur. J. Biochem. 56, 375 (1975)
12. ALBERTI, K.G.M.M., HOCKADAY, T.D.R.: Blood lactic and pyruvic acids in diabetic coma. Diabetes 21, 350 (1972)
13. WATKINS, P.J., SMITH, J.S., FITZGERALD, M.G., MALINS, J.M.: Lactic acidosis in diabetes. Br. Med. J. 1, 744 (1969)
14. DRESCHFELD, J.: The bradshawe lecture on diabetic coma. Br. med. J. 2, 358 (1886)
15. LEXOW, P.: Diabetic coma without ketosis. Acta med. Scand. 163, 115 (1959)
16. DAUGHADAY, W.H., LIPICKY, R.J., RASINSKI, D.C.: Lactic acidosis as a cause of nonketotic acidosis in diabetic patients. N. Eng. J. Med. 267 1010 (1962)
17. WATERS, W.C., HALL, J.D., SCHWARTZ, W.B.: Spontaneous lactic acidosis. Am. J. Med. 35, 781 (1963)
18. COHEN, R.A., WOODS, H.F.: Clinical and biochemical aspects of lactic acidosis. Oxford: Blackwell (1976)
19. WISE, P.H., CHAPMAN, M., THOMAS, D.W., CLARKSON, A.R., HARDING, P.E., EDWARDS, J.B.: Phenformin and lactic acidosis. Br. Med. J. 1, 70 (1976)
20. STAUB, H.: Experimentelle Untersuchungen über Synthalin-Wirkung. Z. Klin. Med. 107, 607 (1928)
21. TRANQUADA, R.E., BERNSTEIN, S., MARTIN, H.E.: Irreversible lactic acidosis associated with phenformin therapy. JAMA 184, 43 (1963)
22. ASSAN, R., HEUCLIN, C., GANEVAL, D., BISMUTH, C., GEORGE, J., GIRARD, J.R.: Metformin-induced lactic acidosis in the presence of acute renal failure. Diabetologia 13, 211 (1977)

23. LUFT, D., SCHMÜLLING, R.M., EGGSTEIN, M.: Lactic acidosis in biguanide treated diabetics. A review of 330 cases. Diabetologia 14, 75 (1978)
24. BERGER, W., MEHNERT-ANER, S., MÜLLY, K., HEIERLI, C., RITZ, R.: 10 Fälle von Laktatazidose unter Biguanidtherapie (Buformin und Phenformin). Schweiz. Med. Wochenschr. 106, 1830 (1976)
25. KNATTERRUD, G.L., KLIMT, C.R., OSBORNE, R.K., MEINERT, C.L., MARTIN, D.B., HAWKINS, B.S.: A study of the effects of hypoglycemic agents on vascular complications in patients with adult-onset diabetes. Diabetes [Suppl.] 24, 65 (1975)
26. BERGER, W., MEHNERT-ANER, S., MÜLLY, K., HINDERLING, P.H.: Lactate tolerance in diabetics undergoing biguanide therapy. Excerpta Med. Int. Congr. Ser. 400, 86 (1976)
27. DOAR, J.W.H., CRAMP, D.G.: The effects of obesity and maturityonset diabetes mellitus on L(+)lactic acid metabolism. Clin. Sci. 39, 271 (1970)
28. LLOYD, M.H., ILES, R.A., WALTON, B., HAMILTON, C.A., COHEN, R.D.: Effect of phenformin on gluconeogenesis from lactate and intracellular pH in the isolated perfused Guinea pig liver. Diabetes 24, 618 (1975)
29. KREISBERG, R.A.: Glucose metabolism in normal and obese subjects. Diabetes 17, 481 (1968)
30. HALL, H., RAMACHANDER, G., GLASSMAN, J.M.: Metabolic effects of phenformin - laboratory studies. Tissue distribution and excretion of phenformin in normal and diabetic animals. Ann. N-Y. Acad. Sci. 148, 601 (1968)

Discussion

C. Perret: How do you explain the disparity of low pH values and only mildly decreased pCO_2 values in your series of patients with biguanide-induced lactic acidosis?

Answer: In most of the patients consciousness was disturbed, and the inadequate ventilatory response to acidosis has to be explained by impaired cerebral function. In addition, three patients were in respiratory failure due to congestive heart failure.

P.A. Poole-Wilson: Is it conceivable that the lower incidence of lactic acidosis during metformin therapy is due to increased awareness of this complication after introduction of the drug? And a second question: What was the ultimate cause of death in your series of patients with lactic acidosis?

Answer: As to your first question - I think that the data from France contradict this suspicion, since metformin was introduced and widely used before this dangerous complication was widely recognised. The cause of death in most of our patients was cardiac failure associated with hypotension and elevated CVP readings.

L. Opie: To what extent does the stimulatory effect of phenformin on glycolysis, as previously reported, contribute to the elevation of blood lactate levels during biguanide therapy?

Answer: You are right; it has been suggested that inhibition of mitochondrial respiration, and thus enhanced glycolytic rates, are important effects of biguanides when relatively high biguanide concentrations are employed. However, stimulation of glucose uptake into skeletal muscle and inhibition of gluconeogenesis in isolated hepatocytes have now been observed in vitro at concentrations too low to be associated with inhibition of mitochondrial respiration (Review in: Davidoff F., Guanidine derivatives in medicine. N. Engl. J. Med. 289:141, 1973).

Chapter 18

Lactate and Pyruvate After Short Exposure to Altitude

J. Weber and P.R. Moret

Respiratory, haemodynamic, and metabolic reactions were studied in 12 normal subjects aged 30-60 years and in nine heart patients aged 50-70 years (these patients had valvular and coronary diseases, but were not in heart failure) at 400 m and 730 mmHg and at 2000 m and 495 mmHg. Immediately after arrival, each subject was examined in a sitting position at rest and with increasing workloads up to submaximal effort on bicycle ergometers. It is interesting to note that the main change at altitude was an increasing cardiac output; the heart rate and the systolic and diastolic blood pressure were generally higher in heart patients at rest and after effort, with no statistical difference following a moderate change of altitude.

Arterial acid-base balance, lactate, and pyruvate were determined. Higher values of lactate were found in heart patients at submaximal effort than in normal subjects on low land and at altitude, with no significant difference between these two altitudes. The values recorded in heart patients are individually plotted and compared with the mean values obtained in normal subjects in Fig. 1.

The variety of heart disease did not allow a statistical analysis of the results. At the highest submaximal workload we found a slight difference, with an average value of 48 mg% at altitude compared with 37 mg% on low land in normal subjects.

We expected to find no difference in anaerobic glycogenolysis in our experiment, the atmospheric pO_2 difference being too small (mean pO_2 74 mmHg on low land and 57.4 mmHg at altitude).

The results of experimentation at different altitudes are controversial. Edwards (1) found a decreased lactate accumulation as a result of strenuous exercise in partially adapted subjects at altitude, which was confirmed by Dill (2) and associates in High School champion runners,

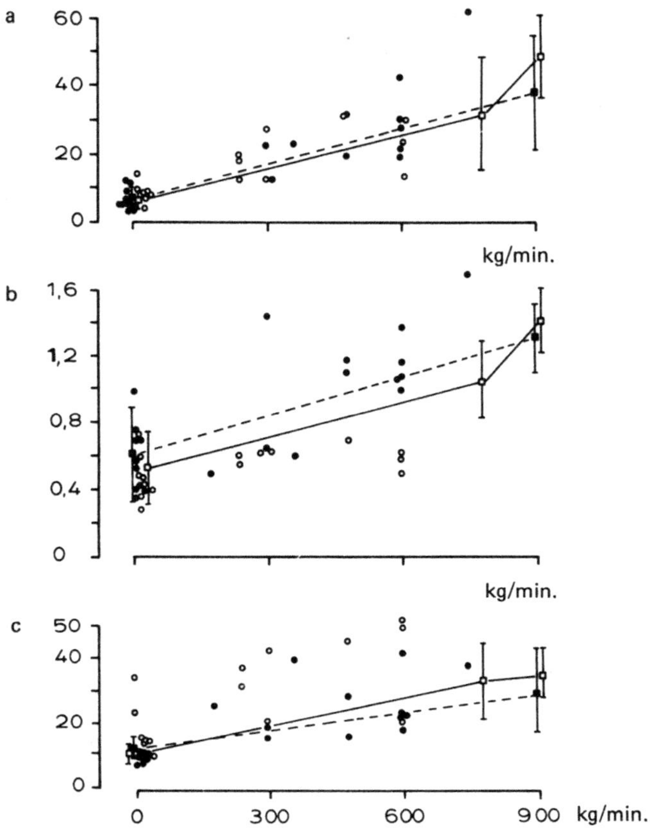

Fig. 1. a–c Lactate (a) and pyruvate (b) levels and lactate:pyruvate ratio (c) in normal subjects on low ground (-■-) and at altitude (-□-) and in heart patients on low ground (●) and at altitude (o)

and by Klausen (3) in three normal subjects. Pugh (4) and collaborators found an elevation of blood lactate during exercise at high altitudes. Hermanson has shown that lactate rises more rapidly after acute exposure to high altitude than at sea level.

It has to be borne in mind that the adaptation periods and the altitudes chosen for experimentation by these different groups are variable.

It can be concluded that in normal subjects and heart patients who are not in heart failure the anaerobic metabolism does not seem to be increased at rest or under average workloads in the adaptation period immediately following a moderate change of altitude.

References

1. EDWARDS, H.T.: Lactic acid in rest and work at high altitude. Am. J. Physiol. 116, 367-375 (1935)
2. DILL, D.B., ADAMS, W.C.: Maximal oxygen uptake at sea level and at 3'090 m altitude in high school champion runners. J. Appl. Physiol. 30, 854-859 (1971)
3. KLAUSEN, K.: Cardiac output in man at rest and during and after acclimatization to 3'800 m. J. Appl. Physiol. 21: 609-616 (1966)
4. PUGH, L.G.C.E.: Cardiac output in muscular exercise at 5'800 m (19'000 ft.). J. Appl. Physiol. 19, 431-440 (1964)

Closing Remarks

C. Perret

Anaerobic glycolysis is an important means of providing energy during heavy muscular exercise continued long enough to deplete the stores of CP. It is interesting to know that in this way athletes develop metabolic acidosis of a similar degree to that measured during diabetic coma. But the surprising thing is the excellent tolerance to such high lactate concentrations and low pH. This supposes adaptive mechanisms, which have to be elucidated.

In heart disease, the delayed adaptation of oxygen transport is responsible for a higher blood lactate concentration for the same submaximal oxygen concentration. But just as in normal subjects, lactataemia varies directly with the fraction of maximal exercise power used.

As expected, the presence of arterial occlusive disease increases the peripheral lactate production during exercise. This explains the abnormally high blood lactate values measured in these conditions during muscular work.

When chronic, hypoxaemia does not necessarily increase blood lactate concentration. This suggests effective compensatory mechanisms. When acute, hypoxaemia can induce moderate hyperlactataemia and acidosis. This rather good tissue tolerance to hypoxaemia contrasts with the important disturbances observed when tissue hypoxia is secondary to lowered perfusion. In circulatory failure and shock, an increased blood lactate concentration may not only be a consequence of a tissue overproduction, but also of a failure of removal. It makes it difficult to use the blood lactate concentration to estimate the extent of tissue anoxia. This also explains the relative value of blood lactate as a prognostic index.

Apart from all these conditions in which accumulation of lactate is related to an oxygen deficit, other situations have been described, associated with severe and sometimes fatal lactic acidosis in absence of clinical signs of anoxia. One main cause is biguanide therapy in diabetic patients with impaired renal function favouring accumulation of the drug. Hyperlactataemia in these conditions appears to be secondary to a decreased uptake rather to an increased production. The precise mechanisms of this metabolic disturbance are not completely understood. They might vary with the type of biguanide concerned and so explain the variable incidence of lactic acidosis observed.

I think that during this session the main problems concerning hyperlactataemia in clinical medicine have been approached. Although many questions have still not been resolved, it is a matter of satisfaction to consider the wonderful advances that have been realised in the understanding of the different mechanisms that can induce an abnormal elevation of lactate in extracellular fluid.

In closing I will thank all of you who participated in the interesting discussions that followed each presentation.

Thank you!

Opening Remarks

P.R. Moret

The last part of our symposium will be specifically dedicated to myocardial ischaemia. The first paper, to be given by Dr. Poole-Wilson, will deal with a rather fundamental aspect of muscular ischaemia, i.e., the reciprocal effect of pH and lactate on contractility.

The next two papers will be given by Drs. Verdouw and Kugler, and will deal mainly with the value of lactate as a marker for ischaemia and with several other markers, which have been advocated in recent years on the basis of results of animal experiments. Some have turned out to be of little clinical use. In this connection, the measurement of the pH in the coronary sinus and that of K and Pi flow come to mind.

During the last 2 days we have talked a lot about lactate in both physiological and pathological conditions. For the cardiologist whose particular interest is in myocardial lactate, it is important to realise that there are other conditions that affect lactate metabolism besides ischaemia. Some of these conditions are shown in Table 1, which is taken from the work of Waters et al. (1963).

Apart from these conditions that modify the amount of arterial lactate, we now know that the heart can choose between carbohydrates and free fatty acids. The amount of lactate consumed or excreted will then depend on which substance the heart chooses for its metabolism.

In the fasting state or in clinical conditions such as excessive lipolysis or heparin injection, myocardial metabolism may be substantially altered, resulting in an increase in free fatty acids.

The last part of this round table discussion will be dedicated to the protection of the ischaemic myocardium, a topic that is much discussed nowadays, which is advocated by some workers and criticised by others,

Table 1. Classification of causes of elevated blood lactate level Waters

Etiology	Range of lactate concentration (mEq/l)
A. Without excess lactate (Normal L:P ratio)	
1. Infusion of pyruvate	2.0-4.9
2. Infusion of glucose	up to 1.7
3. Infusion of $NaHCO_3$	up to 2.4
4. Glycogen storage disease	3.3-14.4
B. With or without excess lactate (normal or increased L:P ratio)	
1. Hyperventilation	1.87-22
C. With excess lactate (increased L:P ratio)	
1. Exercise	8.0-22
2. Epinephrine	1.3-2.9
3. Cardiovascular insufficiency	
a. Shock	2-35
b. Cardiopulmonary bypass	2-8
4. Acute hypoxemia	3-4
5. Severe anemia	up to 1.9
6. Leukemia	12.2-25
7. Diabetes mellitus	10.2-30.8
8. Phenformin therapy	10-31
9. Ethanol ingestion	up to 8
10. Spontaneous (idiopathic)	3.1-23

and which will certainly be a field of intense research in the coming years. This subject will be dealt with by Dr. Hearse.

References

WATERS, W.C., Ill., HALL, J.D., SCHWARTZ, W.B.: Spontaneous lactic acidosis. The nature of the acid-base disturbance and considerations in diagnosis and management. Am. J. Med. 35, 781 (1963)

Chapter 19

Effects of Acidosis and Weak Acids on the Normal, Hypoxic, and Ischaemic Myocardium

P.A. Poole-Wilson

Effects of Acidosis and of Lactate on the Heart

The effects of acidosis and lactate ions on the heart are physiologically important in two common conditions. The first is severe exercise, in which arterial pH may fall to 6.9 and plasma lactate increase to 20 mmol/litre (see this Symposium). The second is myocardial ischaemia presenting as either angina pectoris or acute myocardial infarction. Tennant, in 1935 (1), attributed the rapid decline of myocardial contractility during ischaemia to an intracellular accumulation of hydrogen (H^+) and lactate ions, but was unable to determine their relative significance. More recent work has shown that intracellular acidosis does occur in ischaemia (see Fig. 2; 2) and is of sufficient severity to account for much of the decline in mechanical performance (3, 4). The importance of other factors, such as intracellular accumulation of lactate and phosphate ions (5), oxygen lack, cyclic AMP (6) and the functional capacity of intracellular organelles, remains unclear.

The effects of acidosis, anions and weak acids on the myocardium have been studied by alteration of the constituents of the extracellular fluid. This experimental procedure has important consequences. For example, the effects of increasing extracellular lactate may be akin to the increase that occurs on severe exercise, but they differ substantially from the accumulation of lactate ions in myocardial ischaemia, since this occurs in the intra- and not the extracellular fluid. Similarly, if acidosis is induced by alteration of the extracellular bicarbonate:chloride ratio (metabolic acidosis), the change in pH is largely limited to the extracellular fluid, since H^+ and HCO_3^- cross the cell membrane only slowly (7, 8, 9). A metabolic acidosis, therefore, induces a decrease in the ratio of $[H^+]$ between the intra- and extracellular fluid, which is the exact reverse of what occurs in ischaemia. Alteration of pH by an increase of the extracellular PCO_2 (respiratory acidosis) reduces both intra- and extracellular pH (pH_i and pH_e) and the resulting situation is similar to the predominantly intracellular acidosis of myocardial ischaemia, where it is known that tissue PCO_2 is elevated to as much as 400 mmHg, depending on the experimental conditions (10, 11), and where both pH_i and pH_e are reduced (3, 12).

That an acidosis depresses myocardial function was shown by Klug in 1879 (13) and in 1926 Smith (14) concluded that the fall in contractility was related to changes of pH_i. His conjecture was based on the observation, since confirmed by other workers (15, 16, 17), that the mechanical function of cardiac muscle was reduced more rapidly by a respiratory than a metabolic acidosis. Recent work has shown that pH_i indeed reduced more rapidly by a respiratory acidosis than by a metabolic acidosis (7, 8, 9), probably because H^+ and HCO_3^- cross the cell membrane only slowly, whereas CO_2 is freely diffusible. The cellular mechanism by which pH_i controls function is uncertain and may be attributable to multiple effects (4, 18, 19).

Cardiac muscle responds to acidosis differently from skeletal muscle. Similar changes of pH and PCO_2 in isolated skeletal muscle either do not alter developed tension or cause a small increase (20). The reason for this disparity may be differences in the degree of activation in response to a stimulus (19), in the pH-sensitive dependance of cardiac muscle on extracellular calcium (17, 18), or in the pH sensitivity of the myofibrillar ATPase and sarcoplasmic reticulum (21).

The response of cardiac muscle to an increased concentration of a weak acid in the extracellular fluid depends on several factors. These are (a) the extent to which the free ionised calcium concentration in the extracellular fluid is altered; (b) the rate at which the weak acid, either in the ionised or in the un-ionised form, can cross the cell membrane and alter pH_i; (c) intracellular metabolism of the weak acid accompanied by a production or consumption of protons; (d) activation of a proton pump to correct any change of pH_i that may have occurred; and (e) a direct effect of the weak acid that is not attributable to its acidic properties.

The recent development of calcium electrodes for direct measurement of calcium ion activity has shown that pH changes within the physiological range have little effect on calcium activity in a Krebs solution (22). Replacement of the chloride ion by other ions, however, can alter the activity substantially (23, 24). Experiments performed in collaboration with C.H. Fry and A.J. Williams in our laboratory show that a lactate concentration of 20 mmol/litre in a

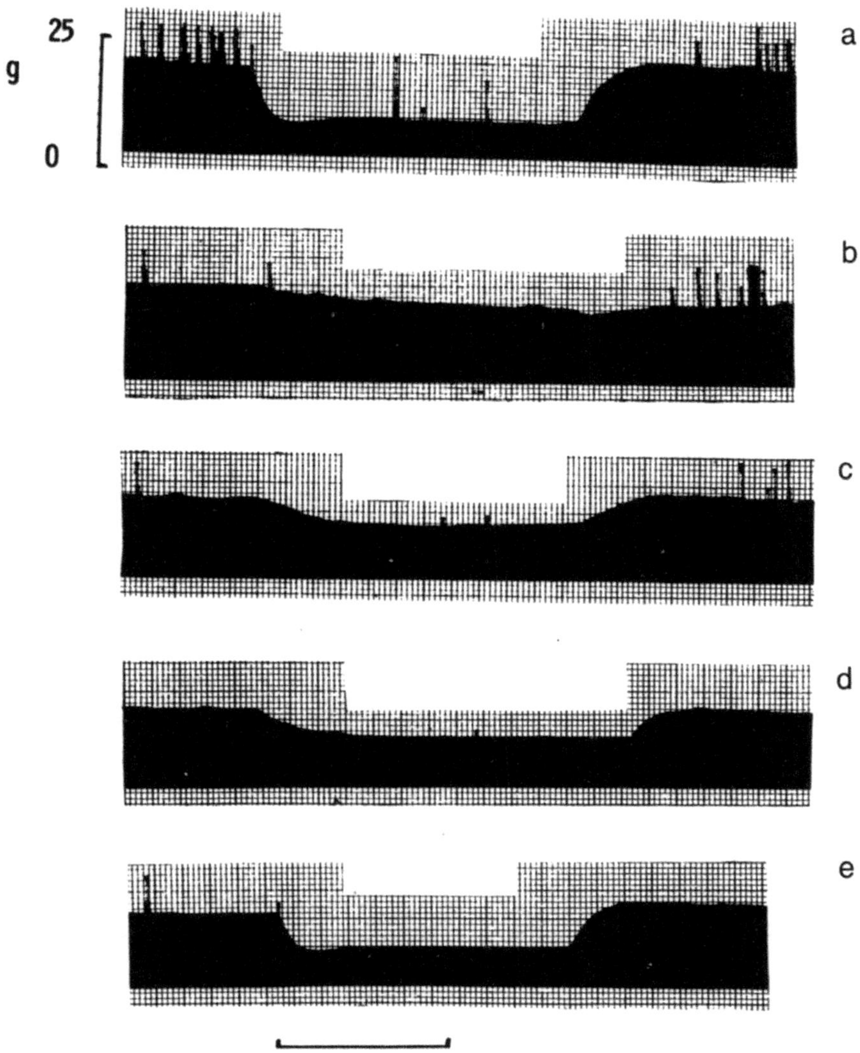

Fig. 1. a-e The effect of several interventions on developed tension in the isolated but arterially perfused interventricular septum of the rabbit. Temperature was 29°C and stimulation rate 48 beats/min. Respiratory acidosis (30% CO_2: a) and DMO (20 mmol/litre: b) cause a rapid fall of developed tension. Addition of lactate ions (25 mmol/litre: c) in exchange for chloride, without any change of pH, causes a similar change in developed tension as a reduction of perfusate [Ca^{2+}]. A reduction of perfusate bicarbonate (metabolic acidosis) to 1.8-0.8 mmol/litre (d) causes a slow fall in developed tension. The pH of the control perfusate was 7.3 (5% CO_2). The pH fell by 0.2 pH units in the presence of 20 mmol/litre DMO (e). Bar indicates 5 min. See text for further description

Krebs solution (pH and osmolality constant, chloride concentration reduced) causes a fall of up to 25% in the calcium ion activity. Thus, the early effect of lactate ions on the developed tension of isolated muscle, as shown in Fig. 1, can be mimicked in both degree and time-course by a reduction of the extracellular calcium (Fig. 1 c and d). This effect of lactate complicates the interpretation of experiments in which changes in the action potential by lactate have been studied (25), since alteration of the free calcium would of itself change the configuration of the action potential. Since such effects can be demonstrated in the extracellular fluid, it is possible that a similar reduction of the free calcium activity in the region of the myofibrils may occur in ischaemia when there is a large increase in the intracellular concentrations of lactate and phosphate (5). Unfortunately, calcium electrodes are not ion-specific for calcium in micromolar concentrations in the presence of millimolar concentrations of sodium and magnesium. It remains possible, therefore, but unproven, that the accumulation of lactate or phosphate ions is a cause of the diminished contractility in myocardial ischaemia.

The effect of the addition of a weak acid to the extracellular fluid on pH_i is dependent on the permeability of the cell membrane to the weak acid, the pK of the weak acid, and pH_e. The myocardial cell membrane appears to be relatively impermeable to H^+ and HCO_3^-, and a metabolic acidosis has only a slow effect on developed tension (Fig. 1; 15). The permeability of cell membranes to the ionised forms of weak acids is, in general, low (26), and weak acids appear to cross the cell membrane mainly in the un-ionised form. Lactic acid has a pK of 3.8 at 25°C and at a pH of 7.40 the ratio of the ionised to the un-ionised form will be approximately 4000:1. If the cell membrane is more permeable by a factor of 10^5 to the un-ionised form, although present in only small amounts, will be the major means by which lactic acid enters the cell and pH_i is altered. Thus, regardless of whether a weak acid or the salt of a weak acid is added to a buffered perfusate, the effect on pH_i is determined by the rate at which the un-ionised form can cross the cell membrane and dissociate in the intracellular fluid; the precise change of pH_i will then be controlled by intracellular buffer mechanisms. The rate at which the un-ionised form of the weak acid crosses the cell membrane will depend on the permeability of the membrane and the external concentration of the un-ionised weak acid which, in turn, varies with pH_e and the pK of the weak acid. Smith (14) has shown that the response of the turtle myocardium in the presence of several weak acids can be explained in this manner. In his experiments there can have been no reduction in the free calcium activity, since he used a Ringer and not a Krebs solution, made from the sodium salt of the weak acid prior to altering pH_e, and pH is known to have little effect on calcium activity (22). In Fig. 1 the effect of lactate can be interpreted as a rapid decline in developed tension due to alteration of the free calcium activity of the perfusate. There is no subsequent decline in developed tension, but the rate of change of developed tension did fall, possibly because of the slow diffusion into the cell of the un-ionised weak acid, which then dissociated to cause a fall of pH_i. Similar results were obtained by Pannier and Weyne (27). In their preparation, time to peak tension was considerably lengthened and after an initial decline, developed tension was unaltered or even increased, but the rate of change of developed tension continued to fall slowly. Fig. 1e illustrates the effect of the weak acid dimethyloxazolidine-dione (DMO). It does not significantly reduce the free calcium activity of a Krebs solution and has a pK of 6.1. At a pH of 7.4 it is therefore relatively un-ionised (20:1). Only the un-ionised form can permeate through the membrane (26), and it causes a rapid fall in tension, presumably due to a change of pH_i similar to that caused by CO_2 (Fig. 1a). Most of the effects of weak acids on myocardial function can be explained in terms of alteration of pH_i and free calcium activity. It is not yet known whether anions such as lactate have any direct effect on the excitation-contraction process, though the duration of the action potential is shortened by lactate (25) and this phenomenon cannot be explained by changes of pH or free calcium activity.

Acidosis in Myocardial Ischaemia

The mechanisms by which an acidosis may occur in myocardial ischaemia have recently been reviewed (28). That an acidosis is present appears certain (see Fig. 2), but the precise severity of the acidosis and the site of the acidosis within the myocardial cell are unknown. Most methods for the measurement of pH_i are invalid in ischaemia (2, 29) and the exact values reported vary greatly (Fig. 2). The results shown in Fig. 2 were obtained under a wide variety of different experimental conditions, but no consistent relation between the severity of duration of ischaemia and fall of pH_i was found with any of them. The

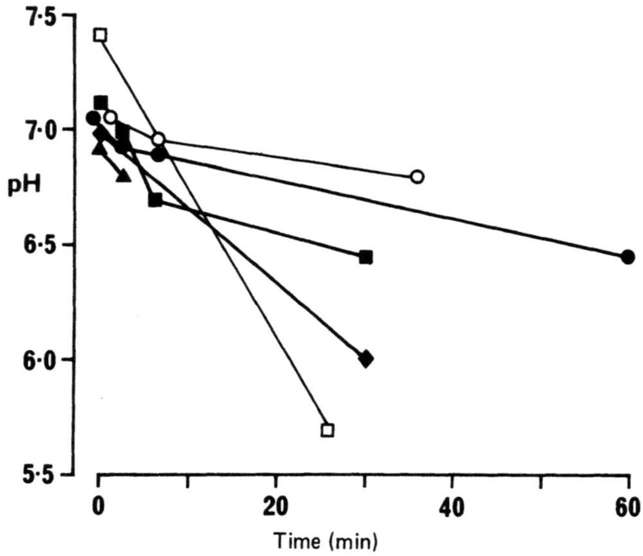

Fig. 2. The values for intracellular pH that have been reported in myocardial ischaemia by Gebert et al. (32) (1971: ■), Neely et al. (12) (1975: o), Krug (41) (1975: ◆), Opie (42) (1976: ●), Hicks et al. (39) (1976: ▲), and Jacobus et al. (40) (1977: □). The experimental preparations and methods for measurement of intracellular pH varied. There is no simple relation amongst these studies between the severity and duration of ischaemia and the fall of intracellular pH

myocardial cell has 40% of its cell volume occupied by mitochondria. Some of the variation in the results may be related to compartmentalisation of H^+ within the cell, which causes problems of interpretation with the DMO or CO_2 methods for measurement of pH_i, and these methods may be invalid in ischaemia for other reasons (2). In the hypoxic myocardium with normal perfusion there is either no change in pH_i or a very slight fall (3, 12, 30, 31). One report suggests a transient alkalosis (32).

A further problem in ischaemia is whether acidosis or lactate accumulation, though diminishing myocardial contractility, is beneficial or deleterious to the eventual recovery of normal function. To assess the role of pH, Poole-Wilson et al. (33) undertook experiments in which hypoxia was induced in an isolated preparation at different pH levels. The results are shown in Fig. 3. As others have reported for a metabolic but not a respiratory acidosis (34, 35), developed tension recovered to a greater extent on reoxygenation if hypoxia occurred concurrent with a reduced pH of 6.8. This beneficial effect of mild acidosis was associated with inhibition of the marked net gain of calcium that occurs on reoxygenation after hypoxia at

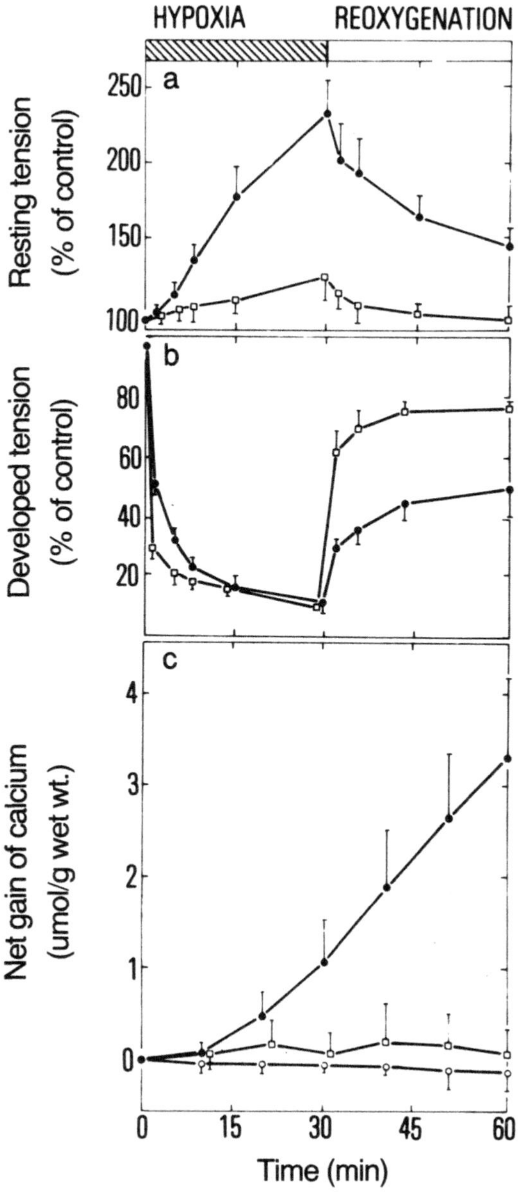

Fig. 3. a–c The effect of respiratory acidosis during hypoxia (▨) on the mechanical function of the isolated but arterially perfused interventricular septum of the rabbit. Temperature was 33°C and stimulation rate 75 beats/min. Resting tension (a) and developed tension (b) are shown. The net gain of calcium is shown in c. Under control conditions (o) there was no net uptake of calcium. Hypoxia was studied with perfusate at a pH of 7.4 (5% CO_2; •; n = 9) and 6.8 (20% CO_2; □; n = 5). Reoxygenation was attained with control perfusate at pH 7.4 (5% CO_2) in all experiments. On reoxygenation after hypoxia at pH 6.8, recovery of mechanical function was increased and net gain of calcium was inhibited. See text for further description

pH 7.4. However, a more severe acidosis (pH 6.6) during hypoxia neither enhanced recovery nor prevented the net gain of calcium on reoxygenation. With a different technique based on use of the isotope $^{47}Ca^{2+}$, Poole-Wilson (2) has shown that a respiratory acidosis does inhibit calcium uptake by the normal myocardium. One, though by no means the only, possible explanation for these results during hypoxia is that acidosis diminishes the influx of calcium into the myocardium on reoxygenation and this prevents intracellular abnormalities of metabolism associated with cytosolic accumulation of calcium. A similar sequence of events may occur in ischaemia (36). Thus, mild acidosis in the myocardium may be partly beneficial. However, if severe, it may hasten cell necrosis by causing either disruption of lysosomes or inhibition of cell metabolism (18).

The acidosis that occurs during severe exercise has more complex and unpredictable effects on the heart. The arterial pH may fall to 6.9 and PCO_2 to approximately 20 mmHg, while the plasma lactate may rise to 20 mmol/litre (see this Symposium). Since CO_2 is rapidly diffusible, the PCO_2 would also fall in the myocardial cell. The resultant increase of pH_i would initially exert a positive inotropic effect (17). Over a period of 1 h, if exercise at such severity could be maintained, H^+ and un-ionised lactic acid would enter the myocardial cell, reduce pH_i, and exert a negative inotropic effect or at least counteract the positive inotropic effect of a low PCO_2. Heart failure might then develop. In addition, a systemic acidosis and exercise are potent stimulants to the release of catecholamines from the adrenal gland, though a severe acidosis inhibits the positive inotropic response of the myocardium to catecholamines (37). Catecholamines also stimulate an H^+ pump in the myocardium, which may delay the onset of a fall of pH_i (38). A further complication will arise if appropriate coronary blood flow is inhibited by atheromatous disease of the coronary arteries or if the energy requirements of the heart exceed its supply of oxygen. Under these circumstances the myocardium is ischaemic, and for that further reason myocardial function would be limited and pH_i would fall.

References

1. TENNANT, R.: Factors concerned in the arrest of contraction in an ischaemic myocardial area. Am. J. Physiol. 113, 677-682 (1935)
2. POOLE-WILSON, P.A.: Measurement of myocardial intracellular pH in pathological states. J. Mol. Cell. Cardiol. To be published.
3. STEENBERGER, C., DELEEUW, G., BARLOW, C., CHANCE, B., WILLIAMSON, J.R.: Heterogeneity of the hypoxic state in perfused rat heart. Circ. Res. 41, 606-615 (1977)
4. POOLE-WILSON, P.A.: Is early decline of cardiac function in ischaemia due to carbon-dioxide retention? Lancet 2, 1285-1287 (1975)
5. KÜBLER, W., KATZ, A.M.: Mechanism of early "pump" failure of the ischemic heart: possible role of adenosine triphosphate depletion and inorganic phosphate accumulation. Am. J. Cardiol. 40, 467-471 (1977)
6. SPERELAKIS, N., SCHNEIDER, J.A.: A metabolic control mechanism for calcium ion influx that may protect the ventricular myocardial cell. Am. J. Cardiol. 37, 1079-1085 (1976)
7. GONZALEZ, N.C., CLANCY, R.L.: Inotropic and intracellular acid-base changes during metabolic acidosis. Am. J. Physiol. 228, 1060-1064 (1975)
8. POOLE-WILSON, P.A., CAMERON, I.R.: Intracellular pH and K^+ of cardiac and skeletal muscle in acidosis and alkalosis. Am. J. Physiol. 229, 1305-1310 (1975)
9. GONZALEZ, N.C., KELLING, E.C., STROME, D.R., CLANCY, R.L.: Cardiac and skeletal muscle acid-base composition during metabolic acidosis in dogs. Respir. Physiol. 26, 1-10 (1976)
10. MacGREGOR, D.C., WILSON, G.L., HOLNESS, D.E., LIXFELD, W., YASUI, H., TANAKA, S., SILVER, M.D., GUNSTENSEN, J.: Intramyocardial carbon dioxide tension: a guide to the safe period of anoxic arrest of the heart. J. Thorac. Cardiovasc. Surg. 68, 101-107 (1974)
11. KHURI, S.F., FLAHERTY, J.T., O'RIORDAN, J.B., PITT, B., BRAWLEY, R.K., DONAHOO, J.S., GOTT, V.L.: Changes in intramyocardial ST segment voltage and gas tensions with regional myocardial ischemia in the dog. Circ. Res. 37, 455-463 (1975)
12. NEELY, J.R., WHITMER, J.T., ROVETTO, M.J.: Effect of coronary blood flow on glycolytic flux and intracellular pH in isolated rat hearts. Circ. Res. 37, 733-741 (1975)
13. KLUG, F.: Über den Einfluß gasartiger Körper auf die Funktion des Froschherzens. Arch. Anat. Physiol. Physiologische Abteilung. 435-478 (1879)
14. SMITH, H.W.: The action of acids on turtle heart muscle with reference to the penetrations of anions. Am. J. Physiol. 76 411-447 (1926)
15. CINGOLANI, H.E., MATTIAZZI, A.R., BLESA, E.S., GONZALEZ, N.C.: Contractility in isolated mammalian heart muscle after acid-base changes. Circ. Res. 26, 269-278 (1970)
16. PANNIER, J.L., LEUSEN, I.: Contraction characteristics of papillary muscle during changes in acid-base composition of the bathing fluid. Arch. Int. Physiol. Biochim. 76, 624-634 (1968)
17. POOLE-WILSON, P.A., LANGER, G.A.: Effect of pH on ionic exchange and function in rat and rabbit myocardium. Am. J. Physiol. 229, 570-581 (1975)
18. WILLIAMSON, J.R., SAFER, B., RICH, T., SCHAFFER, S., KOBAYASHI, K.: Effects of acidosis on myocardial contractility and metabolism. Acta Med. Scand. [Suppl.] 587, 95-111 (1975)
19. TSIEN, R.W.: Possible effects of hydrogen ions in ischemic myocardium. Circulation 53 [Suppl. 1], 14-16 (1976)
20. PANNIER, J.L., WEYNE, J., LEUSEN, I.: Effects of PCO_2, bicarbonate and lactate on the isometric contractions of isolated soleus muscle of the rat. Pfluegers Arch. 320, 120-132 (1970)
21. FABIATO, A., FABIATO, F.: Causes for the more pronounced negative inotropic effect of intracellular acidosis in cardiac than in skeletal muscle. Circulation 56 [Suppl. III] 166 (1977)
22. SCHAER, H.: Decrease in ionized calcium by bicarbonate in physiological solutions. Pfluegers Arch. 347, 249-254 (1974)

23. KENYON, J.L., GIBBONS, W.R.: Effects of low chloride solutions on action potentials of sheep cardiac Purkinje fibers. J. Gen. Physiol. 70, 635-660 (1977)
24. BAND, D.M., FRY, C.H.: An ion selective electrode for the determination of calcium activity. J. Physiol. To be published.
25. WISSNER, S.B.: The effect of excess lactate upon the excitability of the sheep purkinje fiber. J. Electrocardiol. 7, 17-26 (1974)
26. IRVINE, R.O.H., SAUNDERS, S.J., MILNE, M.D., CRAWFORD, M.A.: Gradients of potassium and hydrogen ion in potassium deficient voluntary muscle. Clin. Sci. 20, 1-18 (1960)
27. PANNIER, J.L., WEYNE, J.: The influence of lactate on the contractile properties of papillary heart muscle. Arch. Int. Physiol. Biochim. 78, 101-110 (1970)
28. GEVERS, W.: Generation of protons by metabolic processes in heart cells. J. Mol. Cell. Cardiol. 9, 867-874 (1977)
29. POOLE-WILSON, P.A.: DMO method for intracellular pH. Circ. Res. 39 141-142 (1976)
30. LAI, F., SCHEUER, J.: Early changes in myocardial hypoxia: relations between mechanical function, pH and intracellular compartmental metabolites. J. Mol. Cell. Cardiol. 7, 289-303 (1975)
31. LAVALLÉE, M.: Intracellular pH of rat atrial muscle fibers measured by glass micropipette electrodes. Circ. Res. 15, 185-193 (1964)
32. GEBERT, G., BENZING, H., STROHM, M.: Changes in the interstitial pH of dog myocardium in response to local ischaemia, hypoxia, hyper- and hypocapnia, measured continuously by means of glass microelectrodes. Pfluegers Arch. 329, 72-81 (1971)
33. POOLE-WILSON, P.A., LAKATTA, E.G., NAYLER, W.G.: The effects of acidosis on myocardial function and the uptake of calcium during and after hypoxia. Clin. Sci. Mol. Med. 52, 2-3 (1977)
34. BING, O.H.L., BROOKS, W.W., MESSER, J.V.: Heart muscle viability following hypoxia: protective effect of acidosis. Science 180, 1296-1297 (1973)
35. GREENE, H.L., WEISFELDT, M.L.: Determinants of hypoxia, posthypoxic contracture. Am. J. Physiol. 232, H526-H533 (1977)
36. WEISFELDT, M.L., BISHOP, R.L., GREENE, H.L.: Effects of pH and PCO_3 on performance of ischemic myocardium. In: Recent advances in studies on cardiac structure and metabolism. Roy, P.-E., Rona, G. (eds.), pp. 355-364. Baltimore: University Park Press 1975
37. MITCHELL, J.H., WILDENTHAL, K., JOHNSON, R.L.: The effects of acidbase disturbances on cardiovascular and pulmonary function. Kidney Int. 1, 375-389 (1972)
38. RIEGLE, K.M., CLANCY, R.L.: Effect of norepinephrine on myocardial intracellular hydrogen ion concentration. Am. J. Physiol. 229, 344-349 (1975)
39. HICKS, G.L., HILL, A., DEWEESE, J.A.: Monitoring of midmyocardial and subendocardial pH in normal and ischemic ventricles. J. Thorac. Cardiovasc. Surg. 72, 52-56 (1976)
40. JACOBUS, W.E., TAYLOR, G.J., HOLLIS, D.P., NUNNALLY, R.L.: Phosphorus nuclear magnetic resonance of perfused working rat hearts. Nature 265, 756-758 (1977)
41. KRUG, A.: Alterations in myocardial hydrogen ion concentration after temporary occlusion: a sign of irreversible damage. Am. J. Cardiol. 36, 214-217 (1975)
42. OPIE, L.H.: Effects of regional ischemia on metabolism of glucose and fatty acids: relative rates of aerobic and anaerobic energy production during myocardial infarction and comparison with effects of anoxia. Circ. Res. 38, [Suppl. 1], 52-68 (1976)

Chapter 20

Fundamental Validity and Clinical Usefulness of Myocardial Lactate Balance During Ischaemia A comparison with Other Biochemical Markers

P.D. Verdouw, H. Stam, and W.J. Remme

Introduction

The normally perfused myocardium receives its energy from aerobic pathways. Changes in the amount and distribution of coronary blood flow may lead to changes in oxygen (O_2) and substrate delivery to the myocardium. With an impaired O_2 supply, oxidative phosphorylation is inhibited, so that the energy requirement cannot be met. This leads to the breakdown of high-energy phosphates. Under these conditions the aerobic pathways shift to anaerobic pathways (Pasteur effect), and this shift is accompanied by increased myocardial pyruvate and lactate synthesis and release. The rise in coronary venous lactate concentrations leads to a decrease in the normally positive arterio-coronary venous (A-CV) lactate difference, which may even become negative. Such abnormalities in myocardial lactate balance have been used as an indicator of the presence of myocardial ischaemia in the experimental animal and in man. However, the conditions under which the A-CV lactate difference truly reflects myocardial metabolism must be sharply defined. When one is not aware of the potential pitfalls, the use of lactate levels in the diagnosis of coronary arterial disease may lead to considerable errors. This fact is now generally recognised and has led to the search for other markers of myocardial ischaemia, without the drawbacks of lactate. The purpose of this review is to reconsider the validity and usefulness of lactate measurements during myocardial ischaemia in the light of current knowledge.

Other markers of ischaemia will be compared with lactate and their potential clinical value assessed.

Requirements for Metabolic Markers of Myocardial Ischaemia

Metabolic evidence for myocardial ischaemia obtained in blood samples should not only be sensitive, reproducible, and easy to measure, but must also reflect intracellular metabolism. For the fulfilment of this last requirement the conditions underlying the Fick principle as the basis for metabolic studies must be satisfied (1). When in addition to coronary blood flow (CBF), the arterial concentration ([a]), coronary venous concentration ([cv]), and myocardial uptake or release of a substance are constant, then the equation:

$$\text{Uptake (or release)} = ([a]-[cv]) \times \text{CBF}$$

will hold. From inspection of the equation it becomes clear that A-CV differences are not proportional to tissue metabolism unless CBF is constant, which is not the case in patients undergoing a stress test for the diagnosis of coronary arterial disease. Measurement of CBF in man is still very difficult, and the method generally relies on the thermodilution technique or employs the Fick principle for short periods. Each method determines a mean flow over a period of time and does not provide any information about the redistribution of flow that takes place when the heart becomes ischaemic (2). When regional myocardial flow is not measured, only qualitative information can be obtained from the A-CV differences after a new steady state has been established.

Stress Tests for the Diagnosis of Coronary Arterial Disease

A number of stress tests that may aid in the diagnosis of coronary arterial disease are available. However, not all of them are suitable for the study of myocardial metabolism, although they may provide clear haemodynamic and electrocardiographic evidence of ischaemia. Both exercise and the upper-arm isometric handgrip test release lactate from other points in the body. Infusion of isoproterenol has also been used to provoke ischaemia (3). While arterial lactate does not change significantly during this procedure, released catecholamines may influence the mobilisation of free fatty acids. Here lactate uptake could be decreased due to increased uptake of free fatty acids. With all three test conditions lactate could be a false-positive marker of myocardial ischaemia. Consequently, right atrial pacing at high heart rates is most commonly used, since it is easily executed and reproducible, and costs the patient no subjective effort.

Isometric Handgrip Test

The isometric handgrip effort increases the heart rate and the systemic arterial blood pressure. Consequently the myocardial oxygen requirement will be increased. The haemodynamic effects have been described extensively, and indicate that the test is a useful addition in the diagnosis of coronary arterial disease (4, 5). Recently, two groups have reported on myocardial lactate metabolism during handgrip (6, 7), showing the limited usefulness of lactate as a metabolic marker for ischaemia for the following reasons:

1) The patient is not in a steady state
2) The increase in coronary blood flow during handgrip (20%-30%) is not evenly distributed over the normal and ischaemic zones
3) Arterial lactate increases during handgrip (25%). Handgrip enhances uptake in the normal zone, which may cancel the release by the ischaemic zone if sampling takes place in the coronary sinus only.

Atrial Pacing

Atrial pacing increases the myocardial oxygen requirement through the increase in heart rate. Evaluation of ventricular performance and myocardial metabolism is then possible without the haemodynamic and myocardial consequences of exercise. Unfortunately, there is no internationally accepted protocol according to which heart rates are raised and blood samples are drawn. This complicates comparison of the results of different studies. Some investigators pace the heart at a constant rate during, the entire pacing test (8), although most increase the pacing rate by increments at fixed time intervals (10 beats/min) until angina develops or pacing has to be stopped because of the occurrence of an atrioventricular block.

Lactate as a Marker of Myocardial Ischaemia

Significant increases in coronary sinus lactate are found in 70%-90% of the patients with coronary arterial disease who develop pain during pacing. In general, elevated coronary sinus levels are found before the onset of pain (Fig. 1). This suggests that anaerobic glycolysis occurs very early in the course of ischaemia.

However, a possible role of stress-induced catecholamine release and its metabolic consequences cannot be excluded. Abnormal lactate metabolism is also found in 10%-15% of the patients with proven coronary arterial lesions who do not develop angina during pacing. It could be expected that during anginal attacks all patients would require anaerobic glycolysis in the ischaemic zones. Apparently this is not always reflected in the coronary sinus levels. A frequent cause may be the position of the catheter: the sampling site may be outside the drainage region of the ischaemic area, or the admixture of effluent from the normal zone may dilute the effluent from the ischaemic zone. It is also possible that drainage of the ischaemic

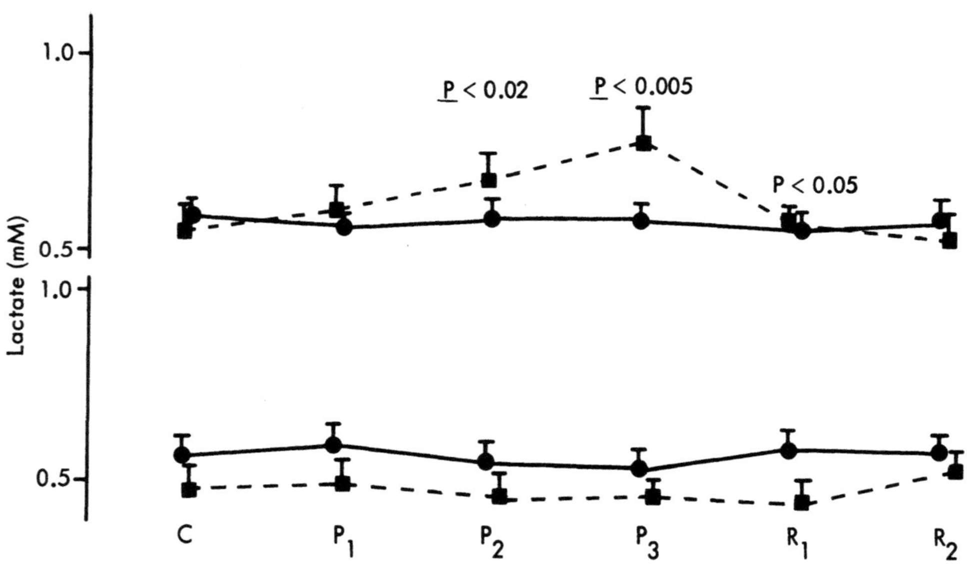

Fig. 1. Arterial (●—●) and coronary (cor) sinus (■—■) levels of lactate during a pacing stress test in 25 patients. Heart rates were raised by increments of 10 beats/ 2 min until pain (anginal group, upper trace, n=18) or an atrioventricular block (nonanginal group, lower trace, n=7) developed. Values given are control values (C), and those recorded in the last three pacing periods (P_1, P_2, and P_3) and after 5 and 10 min of recovery (R_1 and R_2, respectively). Significance levels of differences from control levels are shown

area is not achieved at all. Sampling from multiple sites or local coronary veins therefore offers distinct advantages (3, 9-11). It must be emphasised that problems arising from sampling site and mixing are not limited to lactate alone, but that they will affect all other metabolic markers. Specific problems may arise when the basal state of the patient changes, e.g., when the level of arterial lactate rises after handgrip.

This is especially true for free fatty acids and ketone bodies, which are extracted by the heart in preference to lactate. Thus, when the levels of these substances rise during pacing or any other test, myocardial lactate uptake will be decreased. This could be mistaken for an indication of ischaemia. Metabolic studies should therefore be carried out with the patient in a basal nutritional state. Glycolysis is under the control of glucose transport into the cell and a number of enzymes. Thus factors that affect glucose transport or the activity of these enzymes in addition to ischaemia will also influence lactate release or uptake. A number of clinical conditions have been identified that affect glycolysis. Lack of insulin obviously decreases glucose transport, while elevated free fatty acid levels may also inhibit the passage of glucose to the cytoplasm. The most important point of control of aerobic glycolysis is the activity of phosphofructokinase. Its activity can be increased by factors other than ischaemia (12) (alkalosis and hyperventilation). Table 1 lists a number of factors pertinent to the interpretation of myocardial lactate data (13).

During ischaemia, glycolytic control shifts from phosphofructokinase to glyceraldehyde 3-Phosphate dehydrogenase. Thus, despite an increased activity of phosphofructokinase (by ADP, P_i etc.), glycolysis proved to be diminished, probably due to the inhibition of glyceraldehyde 3-Phosphate dehydrogenase by lactate (14, 15). This can

Table 1. Some disturbing factors in the evaluation of myocardial lactate data

Condition	Mechanism	(a-cs) lactate
Exercise	Art. lactate ↑	↑
Shock	Art. lactate ↑	↑
Diabetes	Insulin ↓ FFA ↑	↓
Starvation	Glycolysis, FFA ↑	↓
Hyperventilation	Phosphofructokinase ↑	↓
Catecholamines	FFA, glycogenolysis ↑	↓
Heparin	FFA ↑	↓

explain the decrease of myocardial lactate release when a coronary artery is occluded for quite a long period of time (10, 16).

It has been assumed that the decrease in lactate release reflects a decrease in ischaemia, due to functioning of a collateral system. However, inhibition of glycolysis is a more plausible explanation. Thus, when the time of onset of a long period of ischaemia is not known (as is often the case during a myocardial infarction), lactate data may prove to be of little help.

An interesting group of patients are those that present with anginal complaints, but who on angiography prove to have normal coronary arteries. In an increasing number of these patients lactate release is found during pacing (17). The clinical relevance of these findings remains uncertain, especially since a large number of these patients show anxiety and have other symptoms that could lead to enhanced glycolysis in the presence of a sufficient oxygen supply. On the other hand, cellular disturbances could exist.

In conclusion, the determination of lactate has proved to be of value as an indicator of cellular dysfunction, but since a large number of factors may modify its release or uptake it is necessary to be aware of possible pitfalls. To increase its sensitivity, lactate has been measured in connection with pyruvate. Henderson et al. (18) demonstrated that pyruvate efflux from isolated rat hearts exceeded that of lactate by a factor of 10. Thus although the lactate:pyruvate ratio may be a more sensitive marker it is also less reliable. In view of these criticisms of lactate data, the usefulness of other markers has been investigated. Some of these markers will be dealt with in the next paragraphs. No attention will be paid to coronary venous pH and O_2 saturation. Although reliable markers in the experimental animal, the results in human patients proved too variable to be of any help in the clinical setting.

Other Markers of Myocardial Ischaemia

Glucose

Restriction of coronary artery flow leads to an increased glucose uptake by the heart (16, 19). In none of the studies could the stoichiometric relationship between glucose and lactate (against uptake of 1 mol glucose there is a release of 2 mol lactate) be demonstrated in the venous blood. Most et al. (20) cited a number of suggested explanations for this: (1) heterogeneity of tissue ischaemia; (2) continued utilisation of endogenous glycogen stores by ischaemic tissue; (3) different transmembrane kinetics for glucose influx and lactate release, regardless of their metabolic relationship; and (4) induction of only partial ischaemia. This would permit hypoxic cells to continue glucose oxidation. Other groups have also shown that glucose determinations during pacing-induced angina (21, 22) are of little clinical relevance.

Potassium

It has long been known that myocardial potassium (K^+) loss occurs during myocardial ischaemia. This has been the subject of various studies, especially because of its possible relationship with the incidence of ventricular arrhythmias (23-27). Studies in dogs (9) and pigs (16) have revealed a sharp rise in coronary venous K^+ immediately after the onset of ischaemia, followed by a gradual return towards arterial levels, but with a second rise after about 30 min. In man, excessive myocardial K^+ release has been reported during pacing-induced ischaemia (28). However, in the same study a smaller but significant K^+ efflux was found in patients who did not become ischaemic during pacing. The authors therefore conclude that tachycardia itself causes a rate-dependent K^+ efflux, and that ischaemia aggravates K^+ losses. Remme et al. (22) only partly confirm these findings. Early rises in coronary sinus K^+ levels occurred at lower pacing rates, but when pacing had to be stopped because of angina or conduction disturbances, no significant arterio-coronary sinus differences could be demonstrated.

ATP Catabolites and Nucleosides

During ischaemia a rapid initial decrease of creatine phosphate (CrP) is observed, while the overall cellular content of ATP falls only moderately (29).

However, it is suggested that the ATP content decreases by different proportions in the various compartments of the myocardial cell. Due to CrP breakdown there will be a rapid accumulation of inorganic phosphate (P_i) in the first minutes of ischaemia. P_i release in the local coronary venous effluent was shown immediately after the onset of myocardial ischaemia in experimental studies (16, 30, 31). However, significant elevations in the overall coronary sinus blood concentration could only be detected when at least 15% of the canine myocardium had become ischaemic (30). P_i concentrations in coronary venous blood draining a specific ischaemic area were considerably more elevated, but decreased after 5-15 min. In these studies P_i release proved much less sensitive than lactate release.

Myocardial P_i release during pacing-induced ischaemia in man is small, and it is significantly elevated in only 55% of anginal patients (32). These results have been confirmed by some other authors (21, 33), whereas Remme et al. (22) found no detectable P_i release. It is concluded that P_i is not a reliable or sensitive marker of ischaemia despite its direct relation to ATP breakdown. The ATP breakdown products, adenosine, inosine, and hypoxanthine, easily pass the cell membrane.

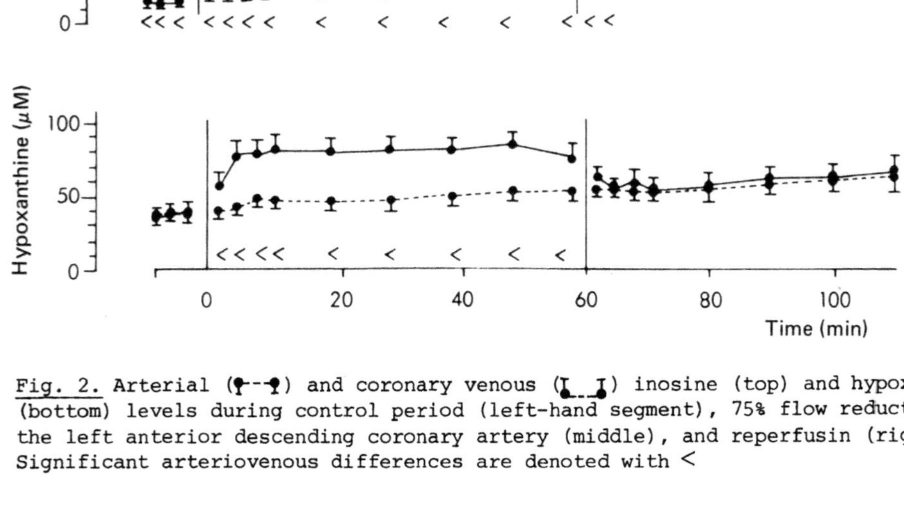

Fig. 2. Arterial (⚲--⚲) and coronary venous (⚲—⚲) inosine (top) and hypoxanthine (bottom) levels during control period (left-hand segment), 75% flow reduction in the left anterior descending coronary artery (middle), and reperfusin (right). Significant arteriovenous differences are denoted with <

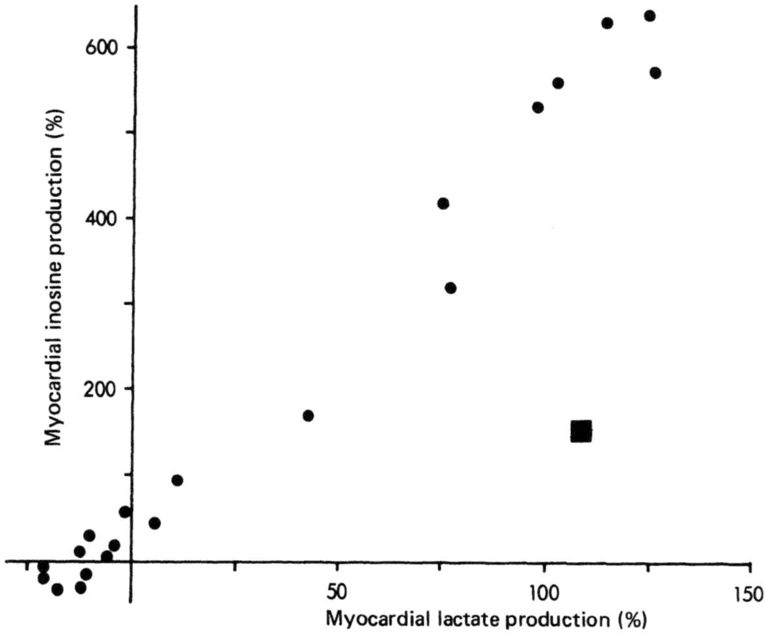

Fig. 3. Relation between myocardial lactate and inosine extraction during control period, ischaemia and reperfusion in the experimental set up described in Fig. 2. ▪ shows the values found after 2 min of ischaemia

Accumulation of the ATP catabolites has been found in the anoxic and ischaemic heart muscle (34, 35). The release of adenosine, inosine, and hypoxanthine has also been found during reactive hyperaemia (36). In our laboratory inosine and hypoxanthine release have been studied in ischaemic porcine hearts (37). (Fig. 2). With the exception of the values found at 2 min of ischaemia, an excellent correlation between lactate and inosine release was demonstrated in controls and during ischaemia and reperfusin (Fig. 3). In the isolated ischaemic rat heart, Stam and de Jong (38) also observed inosine and hypoxanthine release. However, adenosine could not be detected in the coronary effluent, despite appreciable intracellular amounts.

In man, Fox et al. (39) have demonstrated adenosine release during pacing-induced angina. However, the rise in coronary sinus levels was relatively small, probably due to the rapid deamination of adenosine to inosine. On the other hand, appreciable increases in coronary sinus hypoxanthine levels have been demonstrated during angina by several investigators (22, 37, 40, 41). Remme et al. (22) did not find any arterio-coronary sinus differences in hypoxanthine during control conditions, but significant hypoxanthine release occurred during pacing-

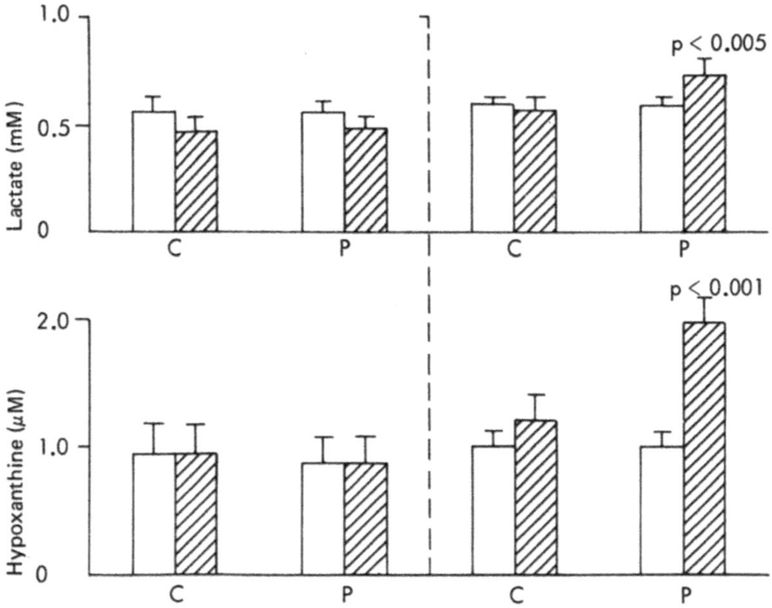

Fig. 4. Arterial (▭) and coronary sinus (▨) levels of lactate and hypoxanthine in anginal (n=17, right-hand panel) and nonanginal (n=7, left-hand panel) patients during control period (C) and the last pacing step (P). Pacing was halted because of pain (in the anginal group) or atrioventricular block (in the nonanginal group). P values refer to arteriovenous differences

induced angina. In this study, hypoxanthine proved to be a more sensitive marker than lactate, but it must be noted that the hypoxanthine levels were very low (Fig. 4). Jorgensen et al. (40) showed that two patients without coronary arterial disease but with the hyperventilation syndrome released lactate but not hypoxanthine. This may indicate that hypoxanthine is potentially a better marker of myocardial ischaemia than lactate. Kugler et al. (41) found hypoxanthine uptake at rest, which decreased during pacing. In the same study inosine uptake was also recorded at rest, which switched to release during angina. This is at variance with the findings of Remme et al. (22), who were not able to detect inosine during pacing. The fact that lactate release is often found before the onset of pain suggests that it is an earlier marker of ischaemia than hypoxanthine.

As argued above, lactate release can, however, also reflect other than ischaemic events (see Section 4). In conclusion, the additional measurement of hypoxanthine in metabolic studies of myocardial ischaemia now seems useful.

Prostaglandins

Since it has been suggested that prostaglandins may play a role in the cardiac response to ischaemia (contractility, vasodilation), their release during experimental ligation of a coronary artery has received considerable attention. (42-44) The site and mechanism of prostaglandin synthesis and release are not completely clear, although evidence has been presented suggesting that synthesis may occur locally in the ventricle (45). Bradykynin, catecholamines, and Ca^{2+} are believed to act on phospholipase A_2, which enhances the formation of arachidonic acid, a precursor fatty acid that is converted into prostaglandins by the action of prostaglandin synthetase.

Berger et al. (46) studied myocardial prostaglandin release during atrial pacing in patients with confirmed multivessel coronary arterial disease. The latter was accompanied by diminished lactate extraction or production. Prostaglandin A was not released during pain and only a small release of prostaglandin E was detectable in the coronary sinus. In contrast, a very significant release of prostaglandin F was found, although no correlation could be established between lactate production and prostaglandin F release. While prostaglandin F has little effect on coronary vascular haemodynamics, it may still have a beneficial effect on the ischaemic myocardium by means of a stabilisation of lysosomes, which could preserve cellular viability. Since prostaglandin release may also be influenced by mediators of stress (47, 48), further investigations are needed to prove their usefulness as markers of myocardial ischaemia. More attention may have to be paid to the recently discovered prostacyclin, a prostaglandin with anti-aggregatory and vasodilatory properties, which might be the major prostaglandin released from the heart (49, 50).

Ketone Bodies

Changes occurring in the redox state of the myocardial cells during hypoxia and ischaemia are reflected in an alteration of the NAD^+: NADH ratio in the different intracellular compartments and can be measured from transitions in redox couples, which are thought to be in equilibrium with the NAD^+-NADH pool.

The lactate-dehydrogenase (LDH) and α-glycerophosphate dehydrogenase (GPDH) systems reflect the cytosolic redox state and the β-hydroxybutyrate-dehydrogenase system (BHBDH) reflects that of the mitochondrial cristae, while the glutamate dehydrogenase system (GDH) represents the redox state in the mitochondrial matrix. Although the β-hydroxybutyrate dehydrogenase is located in the mitochondrial cristae, it only deals with nicotinamide-adenine nucleotides in the mitochondrial matrix. The malate-aspartate shuttle (51) is responsible for the coupling between cytoplasmic and intramitochondrial NAD^+-NADH levels.

The ketone bodies (β-hydroxybutyrate [BHB] and acetoacetate [AcAc]) formed in the liver are suitable substrates for oxidative metabolism in heart (52). The metabolism of these compounds is coupled to the mitochondrial NAD^+:NADH ratios, and alterations may be expected in states of impaired tissue oxygenation. Studies during acute ischaemia in the dog heart showed that the uptake of BHB and AcAc was reduced and that the shift in the mitochondrial redox state was manifested by a rise in the BHB:AcAc ratio in blood sampled from the local coronary vein draining the ischaemic area (53, 54). However, because of the low activity of the BHBDH system, the observed changes in ratio were very small, and not significantly different from simultaneous arterial or coronary sinus values. Furthermore, it was concluded from studies with isolated, perfused rabbit hearts that the BHB:AcAc ratio in the perfusate did not equilibrate with the tissue ratios during control and hypoxic perfusions, as was the case for lactate:pyruvate ratios. Anoxia was accompanied by a reduced AcAc uptake and a shift from BHB uptake to release. For these reasons it can be stated that changes in coronary sinus or local coronary venous ratios of BHB/AcAc have a limited value as markers for ischaemia. A better marker might be the decrease in BHB uptake that takes place during ischaemia.

Amino Acids

Of the amino acids present in the circulation, the branched-chain amino acids (leucine, isoleucine, and valine) and, to a lesser degree, glutamate are taken up by skeletal muscle, while in the heart glutamate uptake is the most important. Alanine and glutamine are released from both tissues. Further evidence indicates a central metabolic role for alanine and glutamine in several oxygen-deprived tissues (55). It has been demonstrated that when oxidative metabolism is depressed, alanine is accumulated and glutamate is depleted in mammalian tissues. The alterations in amino acid metabolism in humans with ischaemic heart disease have been studied at rest and during pacing (56). In the normal myocardium, alanine was only released during pacing, whereas in the diseased heart, a significant alanine release occurred at rest and during pacing. In the coronary group the glutamate uptake was greater than in the normal group at rest, while during atrial pacing glutamate uptake decreased to zero in the control group but persisted in the ischaemic group. On the other hand, lactate balance was abnormal only during pacing in the anginal group.

The possibility that changes in alanine and glutamate metabolism are due to altered protein degradation or synthesis can be excluded, because it has been shown (for papillary muscle) that hypoxia-induced alanine production was not influenced by the protein-synthesis inhibitor puromycine, while amino acid analysis of myocardial protein hydrolysates failed to demonstrate a high alanine prevalence (57). In the same study, direct links among alanine production, pyruvate levels, and tissue glutamate depletion were shown. The finding that alanine formation could be inhibited by the alanine amino transferase inhibitor cycloserine indicates that alanine synthesis takes place from a glutamate-pyruvate transamination.

The enhanced alanine production (from pyruvate and during hypoxia) may represent an NH_4^+ trap following deamination of ATP and amino acids (58). Such a mechanism might be an important system for shunting lactate derived from glycolysis, and this could lead to a decreased lactate formation. Enhanced glutamate uptake and alanine production could be an adaptation of the myocardium to ischaemia. The potency of amino acids as markers for coronary arterial disease is disputable, since alanine and glutamate participate in multiple biochemical reactions, and blood levels may therefore not only reflect cardiac tissue concentrations.

Therapeutic Interventions and Myocardial Metabolism

Protection of the ischaemic myocardium by means of pharmacological, mechanical, or surgical interventions has become a major goal in cardiac research during recent years. Reliable markers of ischaemia are needed to evaluate the effect of these interventions. Myocardial lactate balance has been studied in patients at rest and during pacing before and within 6 months after coronary bypass surgery (59).

Comparison of data obtained pre- and postoperatively may be complicated by possible progression of coronary arterial sclerosis in vessels that were not bypassed. Thus, sampling in veins draining the ischaemic zone may be mandatory when postoperative studies are carried out a long time (1-2 years) after surgery. There is a growing interest in the evaluation of pharmacological drugs during pacing tests (60, 61). In general, the patient serves as his own control. First the rate at which the angina develops is determined and metabolic measurements are carried out. After recovery the drug is administered and again a pacing test is carried out. An important point here is the duration of the recovery period. It has been suggested that the recovery period should be at least 45 min (61). Otherwise successive pacing tests, with or without the administration of drugs, do not yield reproducible results. A second point for consideration must be the short duration of pacing, since it is not obvious that an initial reduction of lactate release would indicate a beneficial effect of use of the drug over a longer period (27).

Acknowledgements

The authors wish to express their gratitude to Prof. P.G. Hugenholtz and Prof. W.C. Hülsmann for their careful reading of the manuscript. Miss J. v.d. Kleijn and Mr. W. Breeman are thanked for their assistance in the preparation of this manuscript.

Summary

Myocardial lactate metabolism has proved to be useful in the diagnosis of myocardial ischaemia.

However, a number of conditions can be defined under which lactate is unreliable as a marker for myocardial ischaemia. Therefore it is desirable to evaluate other markers. ATP catabolites appear to be the most promising, since they are at least as sensitive a marker for myocardial ischaemia and do not have all the drawbacks of lactate. Unfortunately, they are more difficult to determine routinely than lactate.

The ultimate place of biochemical markers (lactate, hypoxanthine, ect.) in the diagnosis and evaluation of treatment of myocardial ischaemia will also depend on the development of new, preferably noninvasive, techniques suitable for the study of regional ventricular performance.

References

1. ZIERLER, K.L.: Theory of the use of arteriovenous concentration differences for measuring metabolism in steady and non-steady states. J. Clin. Invest. 40, 2111 (1961)
2. SCHMIDT, D.H., WEISS, M.B., CASARELLA, W.J., FOWLER, D.L., SCIACCA, R.R., CANNON, P.J.: Regional myocardial perfusion during atrial pacing in patients with coronary artery disease. Circulation 53, 807 (1976)
3. HERMAN, M.V., ELLIOTT, W.C., GORLIN, R.: An electrocardiographic, anatomic, and metabolic study of zonal myocardial ischemia in coronary heart disease. Circulation 35, 834 (1967)
4. LIND, A.R., McNICOL, G.W.: Circulatory responses to sustained hand-grip contractions performed during other exercise, both rhythmic and static. J. Physiol. 192, 595 (1967)
5. HELFANT, R.H., DeVILLA, M.A., MEISTER, S.G.: Effect of sustained isometric handgrip exercise on left ventricular performance. Circulation 44, 982 (1971)
6. NAKHJAVAN, E.K., NATARAJAN, G., SMITH, A.M., DRATCH, M., GOLDBERG, H.: Myocardial lactate metabolism during isometric hand grip test. Comparison with pacing tachycardia. Br. Heart J. 37, 79 (1975)
7. KATORI, R., MIYAZAWA, K., IKEDA, S., SHIRATO, K., MARAGUCHI, I., HAYASHI, T.: Coronary blood flow and lactate metabolism during isometric hand grip exercise in heart disease. Jpn. Heart J. 17, 742 (1976)
8. PARKER, J.O., CHIONG, M.A., WEST, R.O., CASE, R.B.: Sequential alterations in myocardial lactate metabolism, S-T segments, and left ventricular function during angina induced by atrial pacing. Circulation 40, 113 (1969)
9. OWEN, P., THOMAS, M., YOUNG, V., OPIE, L.: Comparison between metabolic changes in local venous and coronary sinus blood after acute experimental coronary arterial occlusion. Am. J. Cardiol. 25, 562 (1970)
10. OBEID, A., SMULYAN, H., GILBERT, R., EICH, R.H., SYRACUSE, N.Y.: Regional metabolic changes in the myocardium following coronary artery ligation in dogs. Am. Heart J. 83, 189 (1972)
11. CORDAY, E., LANG, T., MEERBAUM, S.: Closed chest model of intracoronary occlusion for study of regional cardiac function. Am. J. Cardiol. 33, 49 (1974)
12. SCHEUER, J., BERRY, M.N.: Effect of alkalosis on glycolysis in the isolated rat heart. Am. J. Physiol. 213, 1147 (1967)
13. BRACHFELD, N.: Characterization of the ischemic process by regional metabolism. Am. J. Cardiol. 37, 467 (1976)
14. ROVETTO, M.J., LAMBERTON, W.F., NEELY, J.R.: Mechanisms of glycolytic inhibition in ischemic rat hearts. Circ. Res. 37, 742 (1975)
15. KATZ, A.M.: Physiology of the heart. New York: Raven Press 1977
16. DE JONG, J.W., VERDOUW, P.D., REMME, W.J.: Myocardial nucleoside and carbohydrate metabolism and hemodynamics during partial occlusion and reperfusion of pig coronary artery. J. Mol. Cell. Cardiol. 9, 297 (1977)

17. BOUDOULAS, H., COBB, T.C., LEIGHTON, R.F., WILT, S.M.: Myocardial lactate production in patients with angina-like chest pain and angiographically normal coronary arteries and left ventricle. Am. J. Cardiol. 34, 501 (1974)
18. HENDERSON, A.H., CRAIG, R.J., GORLIN, R.: Lactate and pyruvate kinetics in isolated perfused rat hearts. Am. J. Cardiol. 217, 1752 (1969)
19. BRACHFELD, N., SCHEUER, J.: Metabolism of glucose by the ischemic dog heart. Am. J. Physiol. 212, 603 (1967)
20. MOST, A.S., GORLIN, R., SOELDNER, J.S.: Glucose extraction by the human myocardium during pacing stress. Circulation 65 92 (1972)
21. DAGENAIS, G.R., MARQUIS, Y., MOISAN, A.: Valeurs du glucose, potassium et phosphate inorganique comme indicateurs métaboliques d'ischémie myocardique chez l'humain. Arch. Mal. Coeur 68, 169 (1975)
22. REMME, W.J., DE JONG, J.W., VERDOUW, P.D.: Effects of pacing-induced myocardial ischemia on hypoxanthine efflux from the human heart. Am. J. Cardiol. 40, 55 (1977)
23. HARRIS, A.S., BISTENI, A., RUSSELL, R.A., BRIGHAM, J.C., FIRESTONE, J.E.: Excitatory factors in ventricular tachycardia resulting from myocardial ischaemia. Potassium a major excitant. Science 119, 200 (1954)
24. REGAN, T.J., HARMAN, M.A., LEHAN, P.H.: Ventricular arrhythmias and K^+ tranfer during myocardial ischemia and intervention with procaineamide, insulin or glucose solution. J. Clin. Invest. 46, 1657 (1967)
25. MARSHALL, R.J., PARRATT, J.R.: Antiarrhythmic, haemodynamic and metabolic effects of 3α-amino-5α-androstan-2β-ol-17-one-hydrochloride in greyhounds following acute coronary artery ligation. Br. J. Pharmacol. 55, 359 (1975)
26. VERDOUW, P.D., SCHAMHARDT, H.C., REMME, W.J., DE JONG, J.W.: Antiarrhythmic, metabolic and hemodynamic effects of ORG 6001 (3α amino-5α-androstan-2β-ol-17O-one-hydrochloride after coronary flow reduction in pigs). J. Pharmacol. Exp. Ther. 204, to be published
27. VERDOUW, P.D., REMME, W.J., HUGENHOLTZ, P.G.: Cardiovascular and antiarrhythmic effects of aprindine (AC1802) during partial occlusion of a coronary artery in the pig. Cardiovasc. Res. 11, 317 (1977)
28. PARKER, J.O., CHIONG, M.A., WEST, R.O., CASE, R.B.: The effect of ischemia and alterations of heart rate on myocardial potassium balance in man. Circulation 62, 205 (1970)
29. GUBDJARNASON, S.: The use of glycolytic metabolism in the assessment of hypoxia in human hearts. Cardiology 57, 35 (1974)
30. OPIE, L.H., THOMAS, M., OWEN, P., SHULMAN, G.: Increased coronary venous inorganic phosphate concentrations during experimental myocardial ischemia. Am. J. Cardiol. 30, 503 (1972)
31. MATHUR, P.P., CASE, R.B.: Phosphate loss during reversible myocardial ischemia. J. Mol. Cell. Cardiol. 5, 375 (1973)
32. CHIONG, M.A., WEST, R., PARKER, J.O.: Myocardial balance of inorganic phosphate and enzymes in man. Circulation 49, 283 (1974)
33. DAGENAIS, G.R., JALBERT, B.: Effect of increased free fatty acids on myocardial oxygen extraction and angina threshold during atrial pacing. Circulation 56, 315 (1977)
34. IMAI, S., RILEY, A.L., BERNE, R.M.: Effect of ischemia on adenine nucleotides in cardiac and skeletal muscle. Circ. Res. 15, 443 (1964)
35. OLSSON, R.A.: Changes in content of purine nucleoside in canine myocardium during coronary occlusion. Circ. Res. 26, 301 (1970)
36. BERNE, R., RUBIO, R.: Factors influencing coronary resistance in the ischemic heart. Circulation 39, 40 [Suppl. 4] 240 (1969)
37. VERDOUW, P.D., DE JONG, J.W., REMME, W.J.: Purine nucleoside and carbohydrate metabolism during myocardial ischemia. In: Der Herzinfarkt. Schettler, G., Horsch, A., Mörl, H., Orth, H., Weizel, A. (eds.), pp. 268-281. Stuttgart, New York: Schattauer 1977
38. STAM, H., DE JONG, J.W.: Sephadex-induced reduction of coronary flow in the isolated rat heart: A model for ischemic heart disease. J. Mol. Cell. Cardiol. 9, 633 (1977)
39. FOX, A.L., REED, G.E., GLASSMAN, E.: Release of adenosine from human hearts during angina induced by rapid atrial pacing. J. Clin. Invest. 53 1447 (1974)

40. JORGENSEN, C.R., GOBEL, F.L., NORDSTROM, L.A., NELSON, R.R., FROM, H.L., SIMON, A.B.: Coronary sinus hypoxanthine concentration as an indicator of myocardial ischemia. Circulation 53, 54, [Suppl. II] II-108 (1976)
41. KUGLER, G., KUPPER, H., BLEIFELD, W.: The influence of nitroglycerine on myocardial release of inosine, hypoxanthine and lactate during pacing induced angina in humans with coronary artery disease. J. Mol. Cell. Cardiol. 9, 26 (1977)
42. KADOWITZ, P.J., CHAPNICK, B.M., HYMAN, A.L., BRODY, M.J., GREENBERG, S.: Influence of prostaglandins on the cardiovascular system. In: Cardiovascular Pharmacology. Antonaccio, M. (ed.), pp. 167-184. New-York: Raven Press 1977
43. ALEXANDER, R.W., KENT, K.M., PISANO, J.J.: Regulation of canine coronary blood flow by endogenously synthesized prostaglandins. J. Clin. Invest. 55, 1174 (1975)
44. BERGER, H.J., ZARET, B.L., SPEROFF, L., COHEN, L.S. WOLFSON, S.: Regional cardiac prostaglandin release during myocardial ischemia in anesthetized dogs. Circ. Res. 38, 566 (1976)
45. LIMAS, C.J., COHN, J.N.: Isolation and properties of myocardial prostaglandin synthetase. Cardiovasc. Res. 7, 623 (1973)
46. BERGER, H.J., ZARET, B.L., SPEROFF, L., COHEN, L.S., WOLFSON, S.: Cardiac prostaglandin release during myocardial ischemia induced by atrial pacing in patients with coronary artery disease. Am. J. Cardiol. 39, 481 (1977)
47. NEEDLEMAN, P.: The synthesis and function of prostaglandins in the heart. Fed. Proc. 35, 2376 (1976)
48. MARKELONIS, G.J., GARBUS, J.: Alterations of intracellular oxidative metabolism as stimuli evoking prostaglandin biosynthesis. Prostaglandins 10, 1087 (1975)
49. MONCADA, S., GRYGLEWSKI, R., BUNTING, S., VANE, J.R.: An enzyme isolated from arteries transforms prostaglandin endoperoxides to an unstable substance that inhibits platelet aggregation. Nature 263, 663 (1976)
50. DE DECKERE, F.A.M., NUGTEREN, D.H., TEN HOOR, F.: Prostacyclin is the major prostaglandin released from the isolated perfused rabbit and rat heart. Nature 12, 160-163 (1977)
51. SAFER, B.: The metabolic significance of the malate-aspartate cycle in heart. Circ. Res. 37, 527 (1975)
52. WILLIAMSON, J.R., KREBS, H.A.: Acetoacetate as fuel of respiration in the perfused rat heart. Biochem. J. 80, 540 (1961)
53. WHEREAT, A.F., CHAN, A.: Effects of hypoxemia and acute coronary occlusion on myocardial metabolism in dog. Am. J. Physiol. 223, 1398 (1972)
54. OPIE, L.H., OWEN, P., THOMAS, M., SAMSON, R.: Coronary sinus lactate measurements in assessment of myocardial ischemia. Am. J. Cardiol. 32, 295 (1973)
55. GAILIS, L., BENMOUYAL, E.: Endogenous alanine, glutamate aspartate, and glutamine in the perfused guinea pig heart: Effect of substrates and cardioactive agents. Can. J. Biochem. 51, 11 (1973)
56. MUDGE, G.H., MILLS, R.M., TAEGTMEYER, H., GORLIN, R., LESCH, M.: Alterations of myocardial amino acid metabolism in chronic ischemic heart disease. J. Clin. Invest. 58, 1185 (1976)
57. PETERSON, M.B., TAEGTMEYER, H., RAGAVAN, V.V., FERGUSON, A.G., LESCH, M., BENT, P.: Stimulation of alanine synthesis by anoxia in isolated rabbit myocardium. Am. J. Cardiol. 39, 313 (1977)
58. ROBERTSON, M.G., BREGER, D., SHELL, W.E.: Alanine production by isolated perfused hypoxic hearts: A potential marker of ATP degradation? Abstracts Circ. 55, [Suppl. III] 226 (1977)
59. CHATTERJEE, K., MATLOFF, J.M., SWAN, H.J.C., GANZ, W., KAUSHIK, V.S., MAGNUSSON, P., HENIS, M.M., FORRESTER, J.S.: Abnormal regional metabolism and mechanical function in patients with ischemic heart disease. Circulation 52, 390 (1975)
60. CHIONG, M.A., WEST, R., PARKER, J.O.: The protective effect of glucose-insulin-potassium on the response to atrial pacing. Circulation 54, 37 (1976)
61. JACKSON, G., ATKINSON, L., ORAM, S.: Improvement of myocardial metabolism in coronary arterial disease by beta-blockade. Br. Heart J. 39, 829 (1977)

Chapter 21

Myocardial Release of Lactate, Inosine, and Hypoxanthine During Atrial Pacing- and Exercise-Induced Angina

G. Kugler

Introduction

Myocardial ischaemia leads to anaerobic glycolysis with lactate release from the myocardium contrasting with lactate utilisation by the normally oxygenated heart (1-3). Lactate production, shown very frequently during pacing-induced angina, is not demonstrable during exertional angina, since metabolic evidence of ischaemia is obscured by elevated arterial lactate concentrations (4 and 5). Anaerobic glycolysis has been shown to be accompanied by a decrease in the myocardial tissue content of the high-energy phosphates adenosine triphosphate and creatine phosphate (6 and 7). In animals, after temporary coronary occlusion both an increase in myocardial tissue content (7-9) and a coronary venous release of the diffusible ATP catabolites adenosine, inosine, and hypoxanthine have been shown (10-12).

In humans, other authors have noted the myocardial release of adenosine (13) and hypoxanthine during pacing-induced angina (14), as we did in our own previous investigations on myocardial inosine and hypoxanthine extraction (15-16). In the present study, the coronary venous efflux of lactate, inosine, and hypoxanthine during pacing-induced angina has been compared with myocardial extraction of the catabolites during exertional angina.

Methods

Coronary venous and arterial lactate, inosine and hypoxanthine were measured at rest and during pacing-induced and exertional angina in ten patients with angiographically confirmed disease of one, two, or three vessels in each case with 90% stenosis of the LAD. Lactate was determined enzymatically by conversion to pyruvate with lactate dehydrogenase and NAD. The purine derivatives were separated by column chromatography after ultrafiltration, adsorption onto charcoal, and cellite gel filtration, and then quantitated enzymatically by conversion to uric acid with nucleoside phosphorylase and xanthine oxidase. The mean recovery was calculated as 50% for inosine ^{14}C and 70% for hypoxanthine ^{14}C, which had initially been added to the plasma samples. The lowest, measurable value within the 99% confidence range was 20 nmol/litre for both compounds.

Results

Angina was induced by a mean paced heart rate of 141 ± 21/min with concomitant ST segment depression of 0.21 ± 0.07 mV in ten patients with coronary arterial disease. At a mean resistance of 115 ± 35 W during supine leg exercise performed after a 30-min recovery period, anginal pain was precipitated in all patients, with concomitant ST segment depression of 0.18 ± 0.13 mV. Coronary venous lactate at rest (720 ± 285 μmol/litre) rose during pacing-induced angina (1185 ± 340 μmol/litre ($P < 0.0025$) and increased five fold during exercise (3550 ± 1625 μmol/litre:$P < 0.0005$). The arterial values at rest (850 ± 330 μmol/litre) remained unchanged during atrial pacing (905 ± 280 μmol/litre) and increased five fold during exercise (4380 ± 1860 μmol/litre:$P < 0.0005$). Myocardial lactate extraction at rest (15 ± 9%) changed to production levels (-34 ± 26%) during pacing ($P < 0.0005$), whereas the extraction values increased during exercise (24 ± 13%) ($P < 0.05$) (Fig. 1).

Coronary venous inosine at rest (575 ± 225 nmol/litre) increased both during pacing (1510 ± 945 nmol/litre: $P < 0.005$) and during exercise (1120 ± 620 nmol/litre:$P < 0.01$). The mean arterial value at rest (885 ± 380 nmol/litre) showed a slight increase during pacing (1065 ± 635 nmol/litre:difference not significant) and exercise (965 ± 565 nmol/ litre:difference not significant). The mean myocardial inosine extraction at rest (33 ± 15%) changed to release values (-41 ± 30%) during pacing ($P < 0.0005$) and exercise (-20 ± 27%) ($P < 0.0005$) (Fig. 2).

The coronary venous hypoxanthine levels at rest (1180 ± 890 nmol/litre) rose during pacing (1640 ± 1075 nmol/litre:not significant) and exercise (1995 ± 970 nmol/litre:$P < 0.005$). The arterial values at

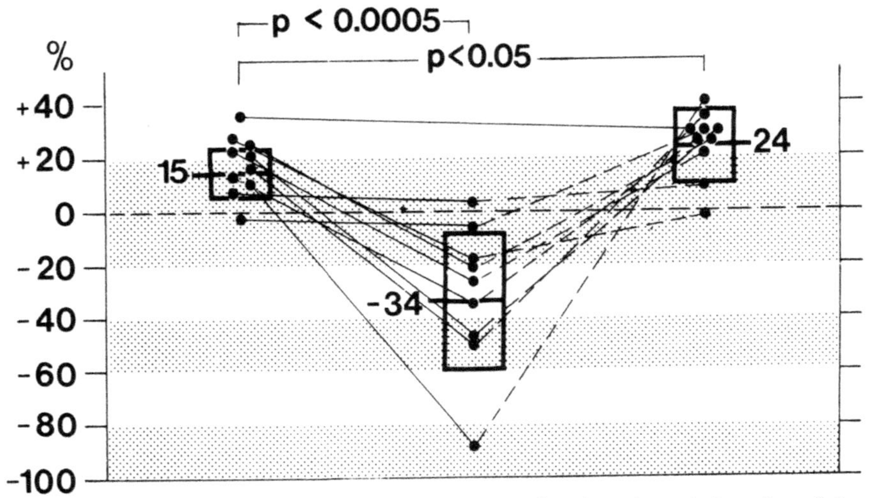

Fig. 1. Myocardial extraction (-) and production (--) levels of lactate at rest (left) and during pacing- (middle) and exercise-induced (right) angina. Significance: Student's paired *t*-test was used

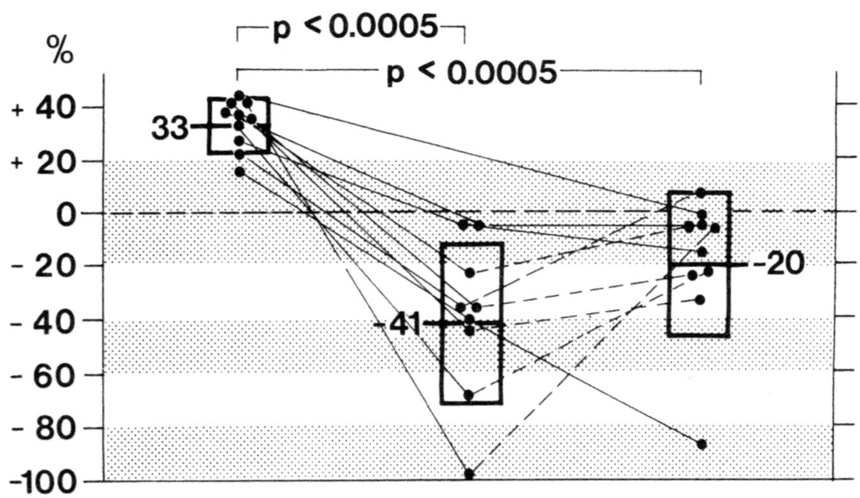

Fig. 2. Myocardial extraction (-) and production (--) levels of inosine at rest (left) and during pacing- (middle) and exercise-induced (right) angina. Significance: Student's paired *t*-test was used

rest (1640 ± 1290 nmol/litre) remained unchanged during pacing (1620 ± 1120 nmol/litre) and increased during exercise (2280 ± 1020 nmol/litre:not significant). The mean hypoxanthine extraction at rest (25 ± 11%) changed to production levels with wide variability during pacing (-7.8 ± 29%) ($P < 0.0025$), and decreased during exercise (10 ± 25%) (not significant: Fig. 3).

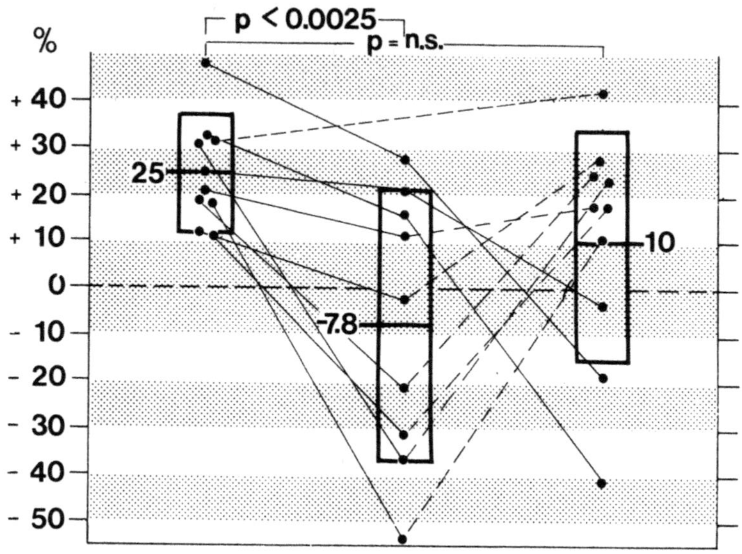

Fig. 3. Myocardial extraction (-) and production (--) levels of hypoxanthine at rest (left) and during pacing- (middle) and exercise-induced (right) angina. Significance: Student's paired t-test was used

Discussion

Pacing-induced myocardial lactate release, which is indicative of anaerobic or aerobic glycolysis, has been used as a qualitative marker of myocardial ischaemia reflecting the cytoplasmatic NADH + H^+:NAD ratio (18-21). The conversion of myocardial lactate extraction to production levels during angina is accompanied by changes in the myocardial inosine and hypoxanthine extraction. Lactate, released from the myocardium during pacing-induced angina, correlates significantly with myocardially produced inosine (15). Thus, the simultaneous release of inosine and lactate functions as a sensitive marker of myocardial ischaemia, since inosine detects ischaemic myocardial energy deficiency by indicating intracellular ATP dephosphorylation and insufficient glycolytic ATP supply. Hypoxanthine extraction, which is significantly reduced during angina, was not sufficiently sensitive in detecting myocardial ischaemia, because of the wide variability. In humans, the reported significant coronary venous release of hypoxanthine (14) was not warranted by the low precision of the assay. The method of Olsson (8) is insensitive for measurements in human blood, because purine concentration are not detectable below 500 nmol/litre, whereas the present complex assay of our own is sensitive to

20 nmol/litre (17). In contrast to simultaneously released inosine and lactate, which are indicative of ischaemic myocardial energy deficiency, unchanged inosine extraction before the onset of pacing-induced angina accounts for sufficient ATP production even if glycolysis is demonstrated with lactate release (15), as shown by Parker et al. (2). Thus simultaneous measurements of lactate and inosine give greater accuracy in the metabolic evaluation of the extent of pacing-induced myocardial ischaemia.

The metabolic evidence of ischaemia during exercise, however, could not be shown by changes of myocardial lactate extraction, because myocardially produced lactate is obscured by elevated arterial concentrations with subsequent wide arteriovenous substrate difference. In agreement with the present study, increased lactate extraction to normal levels during exercise has been reported (4-5) despite the evidence of abnormal left ventricular function. However, significant myocardial inosine release has also been demonstrated, during exertional angina, since arterial inosine values did not alter. Therefore, inosine release indicates that exercise-induced angina is not accompanied by drrminhlx improved myocardial aerobic metabolism such as may be derived from unchanged or increased lactate extraction. In summary, inosine, demonstrating anaerobic metabolism during exertional angina, is an ischaemic marker during pacing-induced and exertional angina.

References

1. HERMAN, M.V., ELLIOT, W.C., GORLIN, R.: An electrocardiographic, anatomic and metabolic study of zonal myocardial ischaemia in coronary heart disease. Circulation 35, 834 (1967)
2. PARKER, J.Q., CHIONG, M.A., WEST, R.Q., CASE, R.B.: Sequential alterations in myocardial lactate metabolism, s-t segments and left ventricular function during angina induced by atrial pacing. Circulation 40, 113 (1969)
3. OWEN, P., THOMAS, M., JOUNG, V., OPIE, L.: Comparison between metabolic changes in local venous and coronary sinus blood after acute experimental coronary arterial occlusion. Am. J. Cardiol. 25, 562 (1970)
4. BOURASSA, U.G., CAMPEAU, L., BOIS, M.A., RICO, O.: Myocardial lactate metabolism at rest and during exercise in ischaemic heart disease. Am. J. Cardiol. 23, 771 (1969)
5. PARKER, J.Q., WEST, R.O., CASE, R.B., CHIONG, M.A.: Temporal relationships of myocardial lactate metabolism, left ventricular function, and s-t-segment depression during angina precipitated by exercise. Circulation 40, 97 (1969)
6. GUDBJARNASON, S.: Acute alterations in energetics of ischaemic heart muscle. Cardiology 56, 232 (1971)
7. KÜBLER, W., SPIECKERMANN, P.G.: Changes in glycolysis and in high-energy phosphates during myocardial ischaemia with intermittent coronary perfusion. Cardiology 56, 100 (1971)

8. OLSSON, R.A.: Changes in content of purine nucleoside in canine myocardium during coronary occlusion. Circ. Res. 26, 301 (1970)
9. OLSSON, R.A., SNOW, J.A., GENTRY, M.K., FRICK, G.P.: Adenosine uptake by canine heart. Circ. Res. 31, 767 (1972)
10. RUBIO, R., BERNE, R.M., KATORI, M.: Release of adenosine in reactive hyperemia of the dog heart. Am. J. Physiol. 216, 56 (1969)
11. DOBSON, J.G., RUBIO, R., BERNE, R.M.: Role of adenosine nucleotides, adenosine and inorganic phosphates in the regulation of skeletal muscle blood flow. Circ. Res. 29, 375 (1971)
12. DE JONG, J.W., GOLDSTEIN, S.: Changes in coronary venous inosine concentration and myocardial wall thickening during regional ischaemia in the pig. Circ. Res. 35, 111 (1974)
13. FOX, A.C., REED, G.E., GLASSMANN, E., KALTMANN, A.J.: Release of adenosine from human hearts during angina induced by rapid atrial pacing. J. Clin. Invest. 53, 1447 (1974)
14. REMME, W.J., DE JONG, J.W., VERDOUW, P.D.: Effects of pacing-induced myocardial ischemia on hypoxanthine efflux from the human heart. Am. J. Cardiol. 40, 55 (1977)
15. KUGLER, G.: Myocardial release of inosine, hypoxanthine and lactate during pacing-induced angina in humans with coronary disease. Eur. J. Cardiol. 9, 227 (1979)
16. KUGLER, G.: The effect of nitroglycerin on myocardial release of inosine, hypoxanthine and lactate during pacing-induced angina. Basic. Res. Cardiol. 73, 523 (1978)
17. KUGLER, G.: A column chromatographic method for determination of plasma and erythrocyte levels of inosine and hypoxanthine. Anal. Biochem. 90, 204 (1978)
18. GORLIN, R.: Assessment of hypoxia in the human heart. Cardiology 57, 24 (1972)
19. OPIE, L.H., OWEN, P., THOMAS, M., SAMSON, R.: Coronary sinus lactate measurements in assessment of myocardial ischaemia. Am. J. Cardiol. 32, 295 (1973)
20. KÜBLER, W.: Myocardial energy metabolism in patients with ischemic heart disease. Basic Res. Cardiol. 69, 105 (1974)
21. BRACHFELD, N.: Characterization of the ischemic process by regional metabolism. Am. J. Cardiol. 37, 467 (1976)

Chapter 22

Protection of the Ischaemic Myocardium

D.J. Hearse

Introduction

Myocardial ischaemia, a condition in which for one of a variety of reasons coronary blood supply is insufficient to meet cellular energy demand, initiates a sequence of cellular changes that, unless halted, will progress through reversible damage and irreversible damage to cell death and tissue necrosis. The rate at which this process of evolving myocardial infarction occurs will be determined by the duration and severity of the ischaemia, and the size of the final infarct will be determined by the number of cells that become irreversibly damaged.

One view of regional ischaemia and evolving myocardial infarction (Fig. 1) is of a central core of severely ischaemic tissue with very low levels of collateral flow, where the cells very rapidly become irreversibly damaged and die. It has been suggested that surrounding this severely ischaemic core, separating it from normally perfused nonischaemic tissue there may be a region of moderately impaired flow, metabolism and electrophysiology. This so-called border zone has been envisaged as containing "jeopardised" cells where the progression of ischaemic damage may be relatively slow, where cellular injury may be of a reversible nature, and where the fate of the cells may lie in a delicate equilibrium for some time.

While the existence of this border zone of jeopardised cells has not been conclusively established (1) and has in fact been questioned (2, 3, 4), it does form a basis for the concept of tissue protection and the reduction of infarct size. According to this concept (5), it should be possible to design interventions to halt or reverse the processes of ischaemic damage in jeopardised cells, where injury is of a reversible nature and where recovery of normal cellular function is at least theoretically possible. The successful use of such interventions should permit the salvage of cells and the reduction of

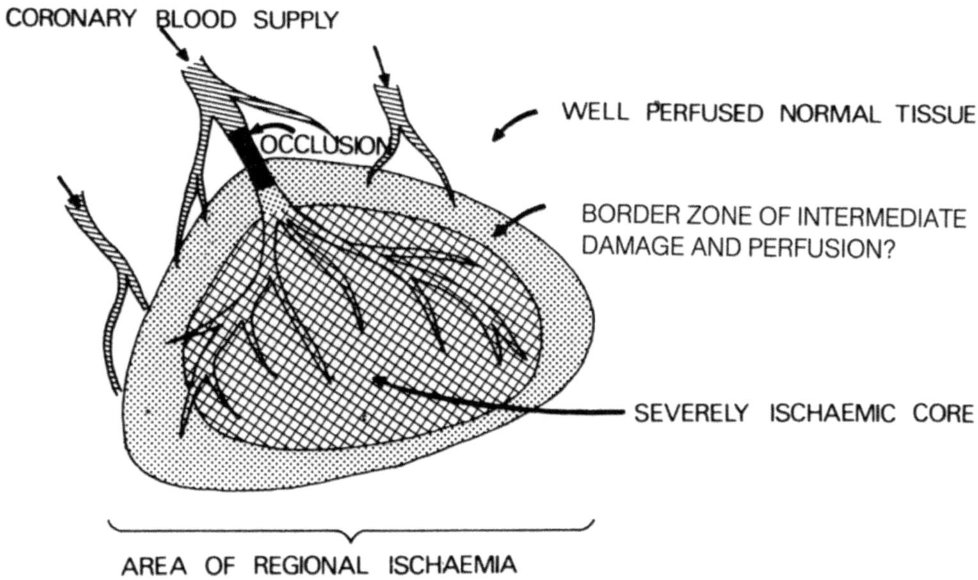

Fig. 1. Regional myocardial ischaemia conventionally visualised as a severely ischaemic central core surrounded by a border zone of tissue exhibiting less severe damage

infarct size. Since infarct size is an important determinant of morbidity as well as mortality, this concept has naturally attracted considerable attention.

In considering approaches to the protection of the ischaemic myocardium, it is perhaps most logical to attempt to characterise the complex continuum of cellular events that follow the onset of ischaemia. Having achieved this, it may be possible to identify critical deleterious changes and then use these changes as focal points for the design of interventions aimed at their reversal.

Figure 2, which is highly speculative, lists the nature and approximate time sequence of some of the important changes known to occur in the myocardium following the onset of severe myocardial ischaemia. The times depicted in Fig. 2 most probably represent the severest ischaemia occurring in the core of the infarct, and the time scale for events occurring in moderately perfused border zone tissue should be correspondingly adjusted. However, the principle remains the same and the knowledge of these sequences has stimulated the development of a number of surgical, biochemical, and pharmacological interventions designed to reduce or delay ischaemic damage and in particular to prevent the transition from reversible to irreversible tissue injury.

Very many interventions have been proposed for protection of the ischaemic myocardium (6, 7), and for the convenience of discussion these may be divided into three groups (Table 1). Firstly, there are procedures designed to reduce myocardial oxygen or energy demand. Secondly, there are procedures for increasing myocardial oxygen or energy supply, and thirdly, there is a miscellaneous group of procedures.

Reduction of Myocardial Energy or Oxygen Demand

Determinants of myocardial oxygen demand include (7, 8) myocardial tension, i.e., internal work, which in turn is determined by ventricular pressure, intraventricular volume, and myocardial mass; contractile state of the heart; heart rate; and the external work performed by the heart. Any decrease in these or their associated parameters may help to prevent or may even reverse the progression of ischaemic damage.

Table 1. Procedures for protection of the ischaemic myocardium

Reduction of myocardial energy or oxygen demand
Reduction of the contractile state of the heart
Optimisation of afterload
Reduction of heart size
Reduction of fatty acid utilization
Control of calcium fluxes
Increasing myocardial energy of oxygen supply
Increasing coronary flow
Increasing blood oxygen content
Increasing oxygen delivery
Modification of substrate utilisation or metabolism
Miscellaneous interventions
Steroids and membrane-stabilising agents
Control of cellular pH
Osmotic control
Hyaluronidase

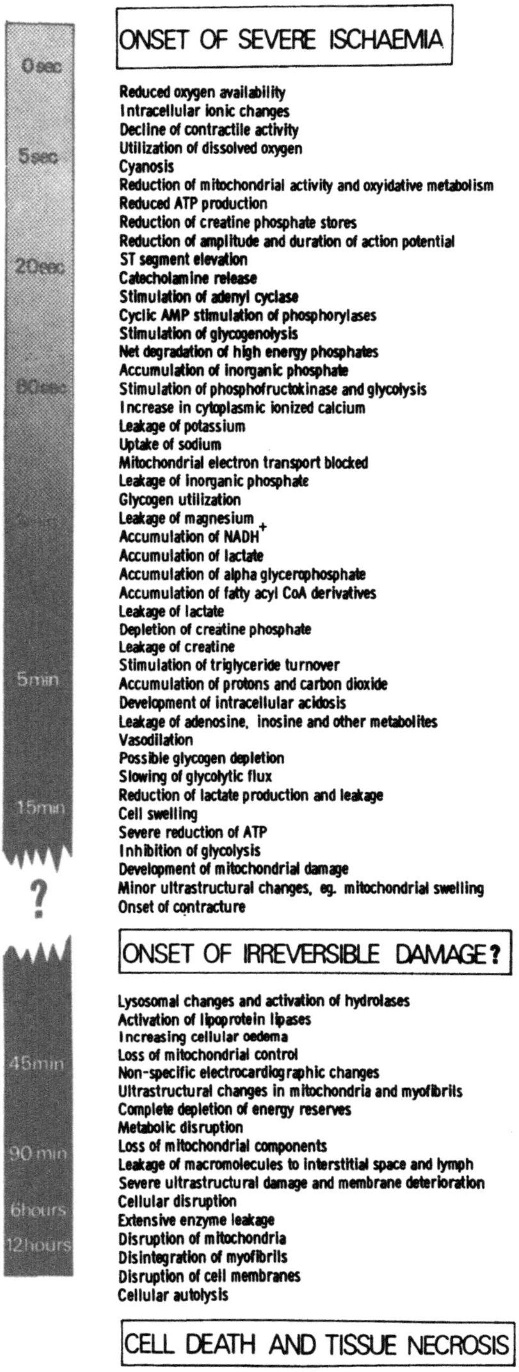

Fig. 2. "Some of the cellular events thought to occur following the onset of severe myocardial ischemia. This figure, which is highly speculative, is not intended to convey that the events occur in strict sequence or in the exact order listed. The progression of changes is a highlydynamic process and only the onset of some changes is indicated. In general the early changes are reversible but with increasing durations of ischemia, progress to irreversible damage. The exact point of the transition is unknown"

Reduction of the Contractile State of the Heart

Beta-adrenergic blockade has been shown (9, 10, 11) to be effective in reducing ischaemic damage through the action of beta blockers in reducing heart rate and inotropic state. In addition, beta blockers have been shown to be effective for the stabilisation of cardiac rhythm. Their use is somewhat restricted, however, since they cannot be used during congestive failure or shock.

Optimisation of Afterload

Haemodynamic interventions for the modification of afterload require complex consideration (6, 12). Reduction of afterload with peripheral vasodilators or intra-aortic balloon counterpulsation may afford protection through the reduction of external work and oxygen demand, but may be detrimental through the associated reduction of arterial pressure and coronary blood flow. Conversely, an increase in afterload may increase oxygen consumption but this may be adequately compensated for by the increased collateral perfusion and tissue oxygen delivery. A delicate balance must therefore be achieved between ventricular function and coronary perfusion. In general, however, it appears that the judicious use of vasodilators, such as hydralazine, nitroglycerin, or nitroprusside (13), or the use of balloon counterpulsation (14, 15) to optimise systemic arterial pressure may reduce myocardial ischaemic damage. However, as will be discussed further in the Section: "Increasing Coronary Flow", although nitroprusside and nitroglycerine both exert similar haemodynamic changes, they may exert opposing effects on myocardial blood flow, nitroprusside inducing a reduction in transmural blood flow (13) with a possible extension of ischaemic damage.

Reduction of Heart Size

Digitalis has been used in the failing heart for the reduction of ischaemic damage (7, 16). While digitalis increases oxygen consumption in the normal heart, in the failing heart the net effect is a reduction of heart size and a slowing of heart rate, with a consequent decrease in wall tension and myocardial oxygen consumption.

Reduction of Fatty Acid Utilisation

Myocardial energy in the form of adenosine triphosphate (ATP) is normally derived predominantly from the oxidation of fatty acids or carbohydrate. The number of molecules of ATP produced per atom of oxygen consumed (the P:O ratio) is lower when fatty acids are used as substrates than when carbohydrates are used. During myocardial ischaemia with a limited availability of oxygen, carhobydrate utilisation represents a more efficient form of energy production. The use of various antilipolytic agents or lipid-free albumin infusions (17) to reduce the level of circulating free fatty acids has been suggested for the reduction of ischaemic damage.

Control of Calcium Fluxes

The role of calcium in contractile activity, its inotropic oxygen-consuming properties, and the implication of calcium overload in the genesis of tissue necrosis (18) have stimulated the search for compounds that limit or control transmembrane calcium movements. A large number of compounds (including beta blockers, antiarrhythmics, barbiturates, and local anaesthetics) with calcium antagonist properties have been described. Perhaps the most interesting are a group (19) of extremely potent calcium antagonist (e.g., verapamil) that can inhibit excitation-contraction coupling in a highly specific way, possibly by the blocking of specific calcium channels. This action, with the consequent depression of ATPase activity, increases cellular energy reserves and thereby reduces the myocardial oxygen demand and may reduce tissue damage during conditions of oxygen deprivation (20, 21).

Increasing Myocardial Energy or Oxygen Supply

Increasing Coronary Flow

Coronary vasodilation with, for example, nitroglycerin can be effective in improving the perfusion of ischaemic tissue and thus reducing myocardial damage (7, 13, 22, 23). The unfavourable lowering of systemic arterial pressure associated with the use of nitroglycerin or other vasodilators can be overcome by the coincident administration of phenylephrine (24). However, when coronary vasodilators are used to increase total coronary blood flow, care must be taken to ensure that there is no "steal" effect with increased coronary perfusion of normal tissue at the expense of the ischaemic tissue. In this connection it has been suggested (13) that while nitroglycerin and nitroprusside cause similar haemodynamic changes, the latter may be harmful through its additional ability to reduce transmural coronary blood flow and hence increase ischaemic damage. Coronary flow and collateral coronary flow may also be increased by drug-induced increases in coronary perfusion pressure, so that the use of inotropic agents such as noradrenalin or dobutamine (25), although increasing heart work and thus oxygen consumption, may exert an overall beneficial effect by a correspondingly larger increase in coronary flow and oxygen delivery. Coronary flow may also be increased by intra-aortic balloon counterpulsation or external counterpulsation. It has also been suggested that coronary flow may be increased by surgical revascularisation. However, this latter approach, although successful in patients with long-term ischaemic heart disease and angina, has attracted considerable criticism when applied to patients with myocardial infarction.

Increasing Blood Oxygen Content

The use of hyperbaric oxygen (26) has been suggested as a means of increasing oxygen delivery without increasing coronary flow. For example, at two atmospheres pressure of pure oxygen the dissolved oxygen content of the blood can be increased by almost 20%. However, hyperbaric techniques are expensive and difficult, and the possibility of oxygen toxicity should not be overlooked. Some success has been reported (27) with the simpler technique of inhalation of oxygen-rich mixtures; however, since haemoglobin is almost fully oxygenated at ambient oxygen tensions only small increases in oxyhaemoglobin are possible.

Increasing Oxygen Delivery

A number of compounds have been shown to be capable of altering haemoglobin-oxyhaemoglobin dissociation curves, and although it has not been tested clinically, it is possible that such an approach may facilitate increased oxygen delivery and the reduction of ischaemic damage.

Modification of Substrate Utilisation or Metabolism

Methods involving modification of the utilisation or metabolism of substrates, which have attracted considerable experimental and clinical interest, are aimed at optimising cellular energy production, either through the removal or provision of certain substrates in the coronary blood or through the manipulation of their metabolism by the activation or inactivation of various enzyme systems. A number of different procedures have been proposed, and these are listed in Table 2.

Glucose, Insulin, and Potassium

Despite the enthusiastic promotion of the use of glucose, insulin, and potassium by Sodi-Pallares and colleagues (28, 29), it has remained at the centre of controversy for many years. Whereas some clinical trials failed to show any improvements in patients treated with this regimen,

Table 2. Biochemical approaches to myocardial protection through the modification of substrate utilisation or metabolism

(i)	Glucose, insulin and potassium
(ii)	Free fatty acids and antilipolytic agents
(iii)	Dichloroacetate and pyruvate dehydrogenase activation
(iv)	Elevation of glycogen reserves
(v)	Extracellular high energy phosphates
(vi)	Inhibition of nucleotide degradation

Table 3. Possible beneficial effects of glucose in the ischaemic myocardium

1.	May stimulate anaerobic ATP production
2.	May reduce ion losses, particularly potassium
3.	May reduce changes in the membrane action potential
4.	May reduce cell swelling
5.	May stabilise cardiac lysosomes
6.	May reduce plasma-free fatty acid concentration

in other trials very striking results have been reported. Recently, the rather decisive British Medical Research Council trial (30) has come under criticism, and it has been suggested (31) that the doses of glucose, insulin, and potassium were too low, the procedure was aimed at preventing potassium loss as opposed to enhancing glucose uptake, and the patient selection was unsatisfactory. Recently there have been several well-controlled animal studies in which glucose has exerted a protective effect in the ischaemic and the anoxic myocardium (32, 33, 34). As Opie (31, 34, 35) and others have pointed out, it is possible to formulate several arguments (Table 3), based both on theory and on experimentation, to support the concept that glucose is beneficial to the ischaemic myocardium. Firstly, by increasing the concentration of extracellular glucose it may be possible to stimulate glycolysis and so improve anaerobic ATP production. This effect may be further enhanced by the coincident administration of insulin. Secondly, it is possible that glucose may prevent the ion losses that occur during ischaemia, particularly the loss of potassium, and here the inclusion of potassium in any treatment may further aid in reducing intracellular potassium loss. Thirdly, it has been shown that glucose is able to reduce ischaemia-induced changes in the transmembrane action potential, and fourthly, while it is always tempting to attribute any action of glucose to its metabolism, it is possible that part of any protective action may lie in its osmotic properties. Fifthly, glucose has been shown to stabilise cardiac lysosomes; and finally, the presence of high concentrations of circulating glucose may act to reduce plasma fatty acid concentrations, which are probably toxic to the ischaemic myocardium.

Free Fatty Acids and Antilipolytic Agents

There is now a convincing body of evidence supporting the concept that high concentrations of free fatty acids or their derivatives are detrimental to the function and the rhythm of the heart (35, 36, 37, 38, 39). Table 4 lists some of these. Firstly, long-chain fatty acyl CoA derivatives have been shown to inhibit adenine nucleotide transferase activity and thus the translocation of energy from the mitochondria to the cytoplasm. Then, fatty acids have been suggested, secondly, to interfere with calcium and magnesium exchange, and thirdly, through their detergent effect, to alter membrane permeability. Fourthly, and in contrast to glucose, fatty acids have been shown to

Table 4. Possible detrimental effects of fatty acids or their derivatives in the ischaemic myocardium

1.	May inhibit adenine nucleotide transferase activity
2.	May interfere with ion exchange, particularly calcium and magnesium
3.	May have a detergent effect upon membrane permeability
4.	May labilise lysosomes
5.	May increase myocardial oxygen consumption
6.	May participate in an energy-wasting cycle of triglyceride synthesis and degradation

labilise lysosomes; and fifthly, fatty acids as substrates, through their reduced P:O ratio, may increase myocardial oxygen consumption. Finally, fatty acids may participate in an energy-wasting cycle of triglyceride synthesis and degradation.

A number of studies (40) have suggested that the use of antilipolytic agents such as β-pyridylcarbinol may reduce ischaemic damage. Other studies (41) have suggested that carnitine could be used to reverse the fatty acyl CoA-mediated inhibition of the adenine nucleotide translocate system.

Dichloroacetate and Pyruvate Dehydrogenase Activation

A number of studies (42) have shown that the compound dichloroacetate can increase cardiac pyruvate oxidation by activating the enzyme pyruvate dehydrogenase from its phosphorylated inactive form. It is possible that the activation of this key enzyme, leading to increased glycolytic flux and decreased fatty acid utilisation, may be beneficial to the ischaemic myocardium, and recent experimental studies (43) in the dog indicate that this may be so.

Elevation of Glycogen Reserves

It is known that glycogen stores afford some protection to the hypoxic myocardium (44). However, under normal conditions myocardial glycogen reserves are small and they can be rapidly exhausted. While there are methods of increasing cardiac glycogen stores, this would need to be carried out in the as yet unidentifiable "preinfarction" state, and even then it appears that the additional protection would be very limited.

Extracellular High-Energy Phosphates

Myocardial ATP, creatine phosphate, and other cellular energy reserves appear to play a critical role in the progression of ischaemic damage, and a number of interventions are specifically aimed at increasing or conserving cellular supplies of these compounds. It is commonly held that cell membranes are impermeable to high-energy phosphates, and if this is so, an extracellular supply of these compounds will have little effect upon their intracellular levels. However, there have now been a number of studies (45, 46) under various experimental conditions, indicating that extracellular ATP or creatine phosphate can afford substantial protection to the hypoxic or ischaemic myocardium. Whether these compounds enter the cell (which may have greatly altered membrane permeability characteristics during ischaemia) or whether they act at the level of the cell membrane is unknown, but their protective effects clearly exist and clearly warrant further investigation.

Inhibition of Nucleotide Degradation

As discussed in the preceding paragraph, a number of interventions have been designed to conserve cellular high-energy phosphates. The net degradation of nucleotides during ischaemia leads to the cellular loss of their degradation products (adenosine, inosine, hypoxanthine, etc.), a situation that may impair subsequent nucleotide resynthesis and tissue recovery. A number of agents have been suggested for the reduction or inhibition of adenine nucleotide degradation. Allopurinol, for example, which inhibits the enzyme xanthine oxidase, has been shown (47) to reduce one facet of ischaemic damage, namely the no-reflow phenomenon, which may limit collateral perfusion or subsequent tissue recovery. Since, the assertion that xanthine oxidase occurs in the myocardial cell has been questioned, the mechanism for the protective action of allopurinol is somewhat abscure, but it may well be that it exerts its action directly upon the epithelial cells of the coronary vasculature. It has been suggested that another compound, dipyridamol (48), reduces myocardial adenine nucleotide degradation by as much as 80% and also induces coronary vasodilation. However, the clinical efficacy of this compound is in question.

Miscellaneous Interventions

A number of procedures that are not aimed directly at energy conservation or energy production have been suggested as effective in the reduction of myocardial ischaemic damage.

Steroids and Membrane-Stabilising Agents

The use of steroids such as methylprednisolone or hydrocortisone, with alleged membrane-stabilising properties, has attracted considerable attention. In general, the results have been conflicting, some studies claiming marked protection while others claim detrimental effects (6, 49, 50). Steroids can undoubtedly be shown to have an effect upon cells and their membranes, and there are a number of studies in which enzyme leakage has been reduced and lysosomal latency and sedimentability have been improved. However, this highlights the importance of determining whether these steroids (or any other potential protective agent under investigation) have any effect upon cell viability, since the reduction of major enzyme release or the maintenance of lysosomal latency, which may possibly be associated with the irreversible damage, may not necessarily be indicative of tissue salvage.

Control of Cellular pH

Disturbances of intracellular pH have been implicated in a number of aspects of cellular damage, and the control of pH by respiratory or metabolic means has been suggested as an effective way of reducing damage during ischaemia or improving postischaemic recovery. In studies in the hypoxic rat heart, Scheuer (51) suggested that either a respiratory alkalosis of pH 7.8 induced with lowered pCO_2 or a similar metabolic acidosis induced with bicarbonate can improve ventricular function during the period of oxygen deprivation and also improve recovery upon reperfusion. In other experiments, Poole-Wilson (52) has shown that a moderate respiratory acidosis induced during hypoxia with a raised pCO_2 can improve posthypoxic function. Clearly the effects of pH are complex and imcompletely understood, but they undoubtedly warrant further investigation.

Osmotic Control

As Willerson, Leaf and others (53, 54) have stressed, osmotic factors can be very important, because ischaemic tissue rapidly becomes oedematous, where upon cellular and subcellular swelling occurs, and this can rapidly lead to myocardial damage. In addition to the effect of cell swelling on the firmness of the ventricular wall and on myocardial membrane tensions, the effect of cell swelling on capillary endothelial cells can be critical. The swelling of these cells may cause the trapping of blood elements, thereby further reducing blood flow and introducing a secondary ischaemia, which may potentiate the initial damage. Control of cellular swelling can be attempted with the use of hyperosmolar solutions, and the results of many studies have suggested that if given by infusion before and during ischaemia, the nonmetabolisable sugar mannitol can improve ventricular function, lessen ST segment elevation, and increase total collateral coronary flow (53). However, the efficacy of mannitol for the reduction of ischaemic injury has recently been questioned and reassessed by Willerson and colleagues (55), and additional experimentation is now clearly required.

Use of Hyaluronidase

Hyaluronidase is a potent enzyme that is capable of digesting mucopolysaccharides. Experimenters working in dog, rat, and human studies have suggested that it reduces myocardial damage and tissue necrosis when given as late as 6 h after the onset of ischaemia (16, 56). The exact mechanism of action of this compound is unknown, but it may act by partially digesting the basement membrane, altering cellular permeability barriers and facilitating substrate availability and metabolite washout. It has been suggested that hyaluronidase improves collateral blood flow to the ischaemic area, but at present it is not clear how this is achieved.

Concluding Comments

A large number of interventions, some of which have been discussed in this brief article, have been proposed for the protection of the ischaemic myocardium. In the case of some interventions there is now strong evidence in support of their efficacy, while in the case of others far more experimental work is required. The evaluation of any intervention is complex, and the modification of a number of indices of damage, e.g., enzyme release of ST segment elevation, as a result of an intervention, does not necessarily justify the assumption that ischaemic damage has been reduced and tissue has been salvaged. Before such protection can be claimed, at least two criteria must be met: firstly, there must be a real improvement in tissue viability, and secondly, tissue damage must be shown to be halted or reduced, and not simply delayed. Convincing evidence of tissue protection should ideally be derived from the measurement of different facets of cellular activity, e.g., conservation of tissue ultrastructure, reduction of enzyme leakage, maintenance of normal metabolic activity and cellular energy reserves, improved contractile activity, and improved electrophysiological characteristics.

Myocardial ischaemia initiates a highly complex sequence of cellular changes; these deleterious changes are not necessarily related and therefore cannot necessarily be combatted with a single intervention. While future investigations in this field will undoubtedly produce new and effective means of reducing tissue injury, perhaps the next major advances will derive from a more broadly based approach in which a number of interventions, each aimed at some different aspect of ischaemic damage, are combined to produce a highly effective intervention.

References

1. HEARSE, D.J., OPIE, L.H., KATZEFF, I.E., LUBBE, W.F., VAN DER WERFF, T.J., PERSACH, M., BOULLE, G.: Characterization of the border zone in acute regional ischemia in the dog. Am. J. Cardiol. 40, 716-726 (1977)
2. CHANCE, B., OPIE, L.H.: The effect of regional ischemia on metabolism of glucose and fatty acids. Circ. Res. 38 [Suppl. I], 52-74 (1976)
3. BARLOW, C.H., CHANCE, B.: Ischemic areas in perfused rat hearts: measurement by NADH fluorescence photography. Science 193, 909-910(1976)
4. MARCUS, M.L., KERBER, R.E., EHDHARAT, J., ABBOUD, F.M.: Three dimensional geometry of acutely ischemic myocardium. Circulation 52, 254-263 (1975)
5. BRAUNWALD, E., MAROKO, P.: The reduction of infarct size - an idea whose time (for testing) has come. Circulation 50, 206-209 (1974)
6. HILLIS, L.D., BRAUNWALD, E.: Myocardial ischemia. N. Engl. J. Med. 296, 971-978, 1034-1041, 1093-1096 (1977)
7. SCHEIDT, S., ALONSO, D.R., WILNER, G., KILLIP, T.: New concepts of cardiogenic shock: preservation of the ischemic myocardium. Bull. N. Y. Acad. Med. 50, 247-260 (1974)
8. SONNENBLICK, E.H., SKELTON, C.L.: Myocardial energetics: basic principles and clinical implications. N. Engl. J. Med. 285, 668-675 (1971)

9. MAROKO, P.R., KJEKSHUS, J.K., SOBEL, B.E., WATANABE, T., COVELL, J.W., ROSS, J.W., BRAUNWALD, E.: Factors influencing infarct size following experimental coronary occlusion. Circulation 43, 67-82 (1971)
10. REIMER, K.A., RASMUSSEN, M.M., JENNINGS, R.B.: Reduction by propranolol of myocardial necrosis following temporary coronary artery occlusion in dogs. Circ. Res. 33, 353-363 (1973)
11. GOLD, H.K., LEINBACH, R.C., MAROKO, P.R.: Propranolol induced reduction of signs of ischemic injury during acute myocardial infarction. Am. J. Cardiol. 38, 689-695 (1976)
12. EPSTEIN, S.E., BORER, J.S., KENT, K.M., REDWOOD, D.R., GOLDSTEIN, R.E., LEVITT, B.: Protection of ischemic myocardium by nitroglycerin: experimental and clinical results. Circulation 53 [Suppl. I], 191-197 (1976)
13. CHIARIELLO, M., GOLD, H.K., LEINBACH, R.C., DAVIS, M.A., MAROKO, P.R.: Comparison between the effects of nitroprusside and nitroglycerine on ischemic injury during acute myocardial infarction. Circulation 54, 766-773 (1976)
14. MUNDTH, E.D.: Mechanical and surgical interventions for the reduction of myocardial ischemia. Circulation 53 [Suppl. I], 176-183 (1976)
15. MAROKO, P.R., BERNSTEIN, E.F., LIBBY, P., DE LARI, G.A., COVELL, J.W., ROSS, J., BRAUNWALD, E.: Effects of intraaortic balloon counterpulsation on the severity of myocardial ischemic injury following acute coronary occlusion: counter-pulsation and myocardial injury. Circulation 45, 552-556 (1974)
16. WATANABE, T., COVELL, J.W., MAROKO, P.R., BRAUNWALD, E., ROSS, J.: The effects of increased arterial pressure and digitalis on the severity of myocardial ischemia in the pharmacologically depressed heart. Ischemia and cardiac depression. Am. J. Cardiol. 30, 371-377 (1972)
17. MJOS, O.: Effect of reduction of myocardial free fatty acid metabolism relative to that of glucose on the ischemic injury during experimental coronary artery occlusion in dogs, experimental and clinical aspects on preservation of the ischemic myocardium. Ed. A. Hjalmarson, L. Werko, Lindgren and Soner, Sweden, Molndal, 1976 pp. 29-34
18. DHALLA, N.S.: Involvement of membrane systems in heart failure due to intra-cellular calcium overload and deficiency. J. Mol. Cell. Cardiol. 8, 661-667 (1976)
19. FLECKENSTEIN, A.: Specific inhibitors and promoters of calcium action in the excitation-contraction coupling of heart muscle and their role in the prevention or production of myocardial lesions. In: Calcium and the Heart. Harris, P., Opie, L.H. (eds.), pp. 135-138. London: Academic Press 1971
20. SMITH, H.J., SINGH, B.N., NISBIT, H.D., NORRIS, R.M.: Effects of verapamil on infarct size following experimental coronary occlusion. Cardiovasc. Res. 9, 569-578 (1975)
21. NAYLER, W.G., GRAN, A., SLADE, A.: A protective effect of verapamil on hypoxic heart muscle. Cardiovasc. Res. 10, 650-662 (1976)
22. SMITH, E.R., REDWOOD, D.R., McCARRON, W.E., EPSTEIN, S.E.: Coronary artery occlusion in the conscious dog: effects of alterations in arterial pressure produced by nitroglycerine, haemorrhage and alpha adrenergic agonists on the degree of myocardial ischemia. Circulation 47, 51-57 (1973)
23. MYERS, R.W., SCHERER, J.L., GOLDSTEIN, R.A.: Effects of nitroglycerine and nitroglycerine-methoxamine during acute myocardial ischemia in dogs with pre-existing multivessel coronary occlusive disease. Circulation 51, 632-640 (1975)
24. BORER, J.S., REDWOOD, D.R., LEVITT, B.: Reduction in myocardial ischemia with nitroglycerine or nitroglycerine plus phenylephrine administered during acute myocardial infarction. N. Engl. J. Med. 293, 1008-1012 (1975)
25. TUTTLE, R.R., POLLOCK, G.D., TODD, G., MacDONALD, B., TUST, R., DUSENBERRY, W.: The effect of dobutamine on cardiac oxygen balance, regional blood flow and infarction severity after coronary artery narrowing in dogs. Circ. Res. 41, 357-364 (1977)
26. ASHFIELD, R., GAVEY, C.J.: Myocardial infarction treated with hyperbaric oxygen. Postgrad. Med. J. 45, 648-654 (1969)
27. MAROKO, P.R., RADVANY, P., BRAUNWALD, E., HALE, S.L.: Reduction of infarct size by oxygen inhalation following acute coronary occlusion. Circulation 52, 360-368 (1975)

28. SODI-PALLARES, D.: The importance of electrolyte therapy in heart disease. Ann. N. Y. Acad. Sci. 56, 603-619 (1969)
29. SODI-PALLARES, D., LEON, J.P., DE BISTENI, A., MEDRANO, G.A.: Potassium, glucose and insulin in myocardial infarction. Lancet 1, 1315-1316 (1969)
30. Medical Research Council: Clinical trial of potassium-glucose-insulin (PGI) treatment in myocardial infarction. Lancet 2, 1355-1360 (1968)
31. OPIE, L.H.: The glucose hypothesis: relation to acute myocardial ischaemia. J. Mol. Cell. Cardiol. 1, 107-115 (1970)
32. MAROKO, P.R., LIBBY, P., SOBEL, B.E., BLOOR, C.M., SYBERS, H.M., SHELL, W.E., COVELL, J.W., BRAUNWALD, E.: The effect of glucose-insulin-potassium infusion on myocardial infarction following experimental coronary artery occlusion. Circulation 45, 1160-1175 (1972)
33. RUSSEL, R.O., ROGERS, W.J., MANTLE, J.A., McDANIEL, H.G., RACKLEY, C.E.: Glucose-insulin-potassium, free fatty acids, and acute myocardial infarction in man. Circulation: 53 [Suppl. I], 207-209 (1976)
34. OPIE, L.H., OWEN, P.: Effect of glucose-insulin-potassium infusion on arteriovenous differences of glucose and of free fatty acids on the tissue metabolic changes in dogs with developing myocardial infarction. Am. J. Cardiol. 38, 310-321 (1976)
35. OPIE, L.H., STUBBS, W.A.: Carbohydrate metabolism in cardiovascular disease. Clin. Endocrinol. Metabol. 5, 703-729 (1976)
36. OPIE, L.H., LUBBE, W.F.: Are free fatty acids arrhythmogenic? J. Mol. Cell. Cardiol. 7, 155-159 (1975)
37. KURIEN, V.A., OLIVER, M.F.: A metabolic cause for arrhythmias during acute myocardial hypoxia. Lancet 1, 813-815 (1970)
38. OLIVER, M.F.: The influence of myocardial metabolism on ischemic damage. Circulation 53, [Suppl. I], 168-170 (1976)
39. SHRAGO, E., SHUG, A.L., SUL, H., BITTAR, N., FOLTS, J.D.: Control of energy production in myocardial ischemia. Circul. Res. 38 [Suppl. I], 75-79 (1976)
40. KJEKSHUS, J.K., MJOS, O.: Effect of inhibition of lipolysis on infarct size after experimental coronary artery occlusion. J. Clin. Invest. 52, 1770-1778 (1973)
41. FOLTS, J.D., SHUG, A.S., KOKE, J.R.: Protection of the ischemic dog myocardium with l-carnitine. Clin. Res. 24, 217A (1976)
42. McALLISTER, A., ALLISON, S.P., RANDLE, P.J.: Effects of dichloroacetate on the metabolism of glucose, pyruvate, acetate, 3-hydroxybutyrate and palmitate in rat diaphragm and heart muscle in vitro and on extraction of lactate, pyruvate and free fatty acids by dog heart in vivo. Biochem. J. 134, 1067-1081 (1973)
43. MJOS, O.D., MILLER, N.E., RIEMERSMA, R.A., OLIVER, M.F.: Effects of dichloroacetate on myocardial substrate extraction, epicardial ST-segment elevation and ventricular blood flow following coronary occlusion in dogs. Cardiovasc. Res. 10, 427-436 (1976)
44. SCHEUER, J., STEZOSKI, S.W.: Protective role of increased myocardial glycogen stores in cardiac anoxia in the rat. Circul. Res. 27, 835-849 (1970)
45. HEARSE, D.J., STEWART, D.A., BRAIMBRIDGE, M.V.: Cellular protection during myocardial ischemia: the development and characterization of a procedure for the induction of reversible cardiac arrest. Circulation 54, 193-202 (1976)
46. PARRATT, J.R., MARSHALL, R.J.: The response of isolated cardiac muscle to acute anoxia: protective effect of adenosine triphosphate and creatine phosphate. J. Pharm. Pharmacol. 26, 427-433 (1974)
47. FABIANI, J.N.: The no-reflow phenomenon following early reperfusion of myocardial infarction and its prevention by various drugs. Heart Bull. 7, 134-142 (1976)
48. KUBLER, W., SPIECKERMANN, P.G., BRETSCHNEIDER, H.J.: Influence of dipyridamol (Persantin) on myocardial adenosine metabolism. J. Mol. Cell. Cardiol. 1, 23-38 (1970)
49. LIBBY, P., MAROKO, P.R., SOBEL, B.E., BLOOR, C.M., COVELL, J.W., BRAUNWALD, E.: Reduction of experimental myocardial infarct size by corticsteroid administration. J. Clin. Invest. 52, 599-607 (1973)
50. ROBERTS, R., deMELLO, V., SOBEL, B.E.: Deleterious effects of methylprednisolone in patients with myocardial infarction. Circulation 53, [Suppl. 1], 204-206 (1976)
51. SCHEUER, J., STEZOSKI, S.W.: The effect of alkalosis upon the mechanical and

metabolic response of the rat heart to hypoxia. J. Mol. Cell. Cardiol. 4, 599-610 (1972)

52. POOLE-WILSON, P.A., LAKATTA, E.G., NAYLER, W.G.: The effects of acidosis on myocardial function and the uptake of calcium during and after hypoxia. Clin. Sci. Mol. Med. 52, 2-3P (1977)
53. WILLERSON, J.T., POWELL, W.J., GUINEY, T.E., STARK, J.J., SANDERS, C.A., LEAF, A.: Improvements in myocardial function and coronary blood flow in ischemic myocardium after mannitol. J. Clin. Invest. 51, 2989-2998 (1972)
54. LEAF, A.: Cell swelling: a factor in ischemic tissue injury. Circulation 48, 455-458 (1973)
55. WILLERSON, J.T., FIXLER, D.E., BUJA, L.M.: Influence of hypertonic mannitol on regional myocardial blood flow and in protecting ischemic myocardium during experimental coronary occlusion. J. Mol. Cell. Cardiol. 9 [Suppl.], 60A (1977)
56. HILLIS, L.D., FISHBEIN, M.C., BRAUNWALD, E., MAROKO, P.R.: The influence of the time interval between coronary artery occlusion and the administration of hyaluronidase on the salvage of ischemic myocardium in dogs. Circul. Res. 41, 26-31 (1977)

Closing Remarks

P.R. Moret

The last part of this symposium, which was specifically dedicated to the question of the ischaemic myocardium, was begun by an excellent paper by Dr. Poole-Wilson on the effects of acidosis and weak acids on the normal, hypoxic, and ischaemic myocardium. By making use of the results of previously published works and his own researches, he attempted to dissociate the effects of acidosis and of lactate ions on myocardial contractility.

Various physiological or pathological conditions, such as strenuous physical exercise, ischaemia, hypoxia and metabolic or respiratory acidosis, are accompanied by changes in pH or lactate, which may initially be more marked in the intracellular space or in the extracellular space. As a result of this, the effects on myocardial contractility are different.

The balance in the ion content between the extracellular and the intracellular space depends to a large extent on the permeability of the cellular membrane to the different ions. This permeability differs for the different ions under consideration.

The results obtained by metabolic and respiratory acidosis or by ischaemia and hypoxia are enough to remind us, as clinicians, that the situation is more complex than we sometimes appreciate. Dr. Poole-Wilson has furthermore reminded us that any experiment designed to study the effects of acidosis or the accumulation of lactate must always consider the manner in which they are produced.

It seems to be accepted by many workers that the intracellular acidosis present in ischaemia plays a part in reducing contractility, and it is probable that the accumulation of lactate and phosphate also contribute by way of modifying the pH or the activity of free Ca^{2+}.

The author also discussed the question as to whether mild acidosis or some accumulation of lactate might not be of benefit for the myocardium during the recovery phase following ischaemia. On the other hand, severe acidosis with a high accumulation of lactate could have a deletarious effect, since it speeds up cellular necrosis.

The papers by Drs. Verdouw and Kugler are of great practical importance for the clinician, since they illustrate the diagnostic values of the different markers of hypoxia for the detection and diagnosis of myocardial ischaemia. Dr. Verdouw should be specially congratulated for his comprehensive and critical review. The most frequently used marker is still lactate. Its disadvantage is, as Dr. Verdouw pointed out, its low sensitivity and specificity when used during an exercise test or the handgrip test (increase in quantity in arterial lactate) or during trials with isoproterenol (increase in free fatty acids, which causes a decrease in lactate consumption). In practice, the use of a lactate test is possible only during pacing, which is not the best way to produce ischaemia. An exercise test, which corresponds more closely to the normal everyday life of our patients, is the ideal way to provoke ischaemia.

The results obtained from animal experiments had given clinicians the hope that many other indicators might be useful. However, in practice, it has been found that measurements of the blood pH, the oxygen content of coronary venous blood, the glucose consumption of the myocardium, and the excretion of potassium are of little clinical use. The measurements of the catabolites of adenosine or of the nucleotides appear to have opened up new paths, in particular hypoxanthine and inosine. The determination of Pi, on which a lot of hope was placed, is not very reliable and also not very sensitive. According to Dr. Verdouw, hypoxanthine could be a better marker than inosine, but unfortunately it is present in blood only in small quantities. Dr. Kugler believes that the variations in hypoxanthine are not sufficienty significant to make it possible to detect ischaemic areas, and prefers inosine. The big advantage of measuring inosine or hypoxanthine is that both of these markers can be applied during exercise tests.

More recent research work on prostaglandins, and in particular prostaglandin F, inspires new hope in this respect, but more clinical studies are needed. Measurements of ketone bodies (beta-hydroxybutirate and acetoacetate) have also found little clinical use. The measurement of some amino acids, such as glutamate, still remain undefined.

In conclusion, one can say that lactate is still a good indicator of myocardial ischaemia. However, one has to be aware of the numerous conditions that can affect its production and utilisation and can lead to false interpretations. These conditions include diabetes mellitus, fasting, hyperventilation, adrenergic hyperactivity, and the concurrent use of heparin. The lactate test can be used only at rest or during pacing, and not during exercise. At present it appears that only measurements of hypoxanthine or inosine and perhaps prostaglandin F can be employed in the detection of myocardial ischaemia during exercise.

The last paper of this symposium was dedicated to the problem of the protection of the ischaemic myocardium. The excellent presentation made by Dr. Hearse summarises the overall picture, including current practice and the possibilities for the future. The concept of the protection of the myocardium rests on two main assumptions - first that there is a "border zone" of viable tissue near the dead myocardium (this idea is not universally accepted), and second that this area can be salvaged. However, most clinicians cannot say precisely how long ischaemia has lasted and how severe it has been.

Among the agents in use at the moment that Dr. Hearse has mentioned and many workers have admitted to be of clinical use are beta-blocker drugs, which reduce the afterload, and digitalis, for patients with large hearts.

Some other drugs are used by some clinicians and questioned by others. These include the drugs reducing lipolysis, the inhibitors or blockers of calcium transport through the membrane (verapamil), solutions based on glucose, insulin and potassium, solutions tending to influence cellular volume (mannitol), and lastly those that act specifically on the cell structure, such as steroids, hyaluronidase or dipyridamol. Substances that have a more direct metabolic action, such as dichloroacetate or piridoxilate, which we have studied (1 and 2) are still in the experimental stage. These substances with metabolic actions and the exact mechanism by which they act certainly deserve more intense research.

Coronary arterial surgery is useful in chronic stable angina. However, during acute ischaemia or infarction, the indications for surgery are limited to such conditions as unstable angina that has become resistant to all medical treatments.

The last paper by Dr. Hearse, in which he purposely struck a note of hope, opens up a wide field of investigation. All the approaches discussed aim at protecting the myocardium during ischaemia. However, it is our duty both as researchers and as clinicians to find ways to prevent ischaemia by eliminating its causes rather than protecting the myocardium when ischaemia is already present.

References

1. MORET, P.R., LUTZEN, U.: Protection du myocarde hypoxique. Ann. Cardiol. Angeiol. (Paris) 26, 85 (1977)
2. MORET, P.R., LUTZEN, U.: Effect protecteur du piridoxilate sur le myocarde hypoxique. Etudes experimentales. Schweiz. Med. Wochenschr. 107, 1585 (1977)

Conclusion

H. Denolin

The meeting in Anzère on anaerobic metabolism and lactates was an excellent one and we have to sincerely thank our colleagues J. Weber, P.R. Moret, and J.C. Haissly for having planned and organised it so successfully.

The lectures were clear and concise and can hardly be summarised, but we will briefly comment on them.

The first communications concerned the fundamental aspects of energy production and the synthesis and metabolism of lactates. The problem of oxygen debt was discussed once more. It is an old problem: in 1942, when I was in Paris, Prof. Y. Bouvrain wrote a thesis on the measurement of the oxygen debt in cardiac patients as a method for clinical evaluation. In 1947 I attempted to measure it in normal individuals working under conditions of hypoxia. Later on I came to the conclusion that the oxygen debt was difficult to measure and hardly reproducible in normal subjects and would probably not be retained as a functional evaluation method in current practice (see also Mohler and Armstrong, Respiratory Physiology, 1973, 17, 248). The discussion on the validity of the term "debt" was valuable, although it is an old subject of discussion (Christensen, 1950). Why should we not maintain the term debt, which is well accepted and says what it means, even if we do not properly understand the reason why this debt is not made up for, or its physiological mechanisms. The conclusion of Dr. Wilkie in this respect should be carefully studied.

The problem of the anaerobic threshold was discussed; it is an interesting, although rather arbitrary concept. We tried to define this threshold by noninvasive methods, by evaluating the ventilatory function during exercise, but it is sometimes difficult to observe a marked change in the ventilation curve or in the evaluation of the respiratory equivalent of oxygen. This problem and the significance of the maximum rate of lactate formation during continued and very intense exercise tests need to be further investigated.

The application of these data to the physiology of sports or to clinical practice in the case of cardiac, vascular, pulmonary, or metabolic diseases - deserves further studies and Dr. Perret has summarized the present status and outlined the problems to be solved very well.

Several speakers discussed myocardial ischaemia and stressed the importance of the concentration of lactates in coronary lesions as a sign of myocardial ischaemia, namely in the interpretation of some clinical conditions in which pain is associated with a normal coronary angiography. Perhaps other aspects of myocardial metabolism that are measurable in man also deserve new investigations, as Professor Moret pointed out.

We concluded with a particularly interesting subject, which is of present-day interest: the protection of the ischaemic heart. Possible solutions were put forward and we can only hope that some of these will be of practical interest and will be applied in the care of coronary patients.

As I pointed out at the beginning of these brief conclusions, the papers presented at this symposium were very carefully prepared and need few additional remarks. I have only attempted to point out that the clinician is, and will more and more become, confronted with the concepts of anaerobic and lactate metabolisms, either at the level of the entire organism or at that of the myocardium.

It was necessary to have an update on these questions, because they had not been broadly discussed for a long time. Naturally there are still many open questions, but new avenues seem to come up in fundamental research as well as in clinical practice. Once again I should like to express our appreciation to the organisers of this symposium, who have made this important meeting possible.

Subject Index

FEBS

European Journal of Biochemistry

ISSN 0014-2956 Title No. 225

The **European Journal of Biochemistry** will primarily publish papers on fundamental aspects of biochemistry, molecular biology of cellular biology provided that they describe results which make a sufficient contribution, either experimental or theoretical, to our understanding of biological problems at the chemical or physical level, or that they describe new methods applicable to biochemical problems.

The Table of Contents of the **Journal** will be subdivided as follows:

1. Nucleic acids, protein synthesis, and molecular genetics.
2. Protein chemistry and structure.
3. Enzymology.
4. Carbohydrates, lipids, and other natural products.
5. Physical biochemistry.
6. Membranes and bioenergetics.
7. Cellular biochemistry and metabolism.
8. Developmental biochemistry and immunology.

Springer-Verlag
Berlin
Heidelberg
New York

Published by Springer-Verlag
Berlin Heidelberg NewYork
On behalf of the Federation of European Biochemical Societies

GPSR Compliance

The European Union's (EU) General Product Safety Regulation (GPSR) is a set of rules that requires consumer products to be safe and our obligations to ensure this.

If you have any concerns about our products, you can contact us on ProductSafety@springernature.com

In case Publisher is established outside the EU, the EU authorized representative is:

Springer Nature Customer Service Center GmbH
Europaplatz 3
69115 Heidelberg, Germany

Batch number: 10584556

Printed by Printforce, the Netherlands